Martin Pollow

Directivity Patterns for Room Acoustical Measurements and Simulations

Logos Verlag Berlin GmbH

Aachener Beiträge zur Technischen Akustik

Editor:
Prof. Dr. rer. nat. Michael Vorländer
Institute of Technical Acoustics
RWTH Aachen University
52056 Aachen
www.akustik.rwth-aachen.de

Bibliographic information published by the Deutsche Nationalbibliothek

The Deutsche Nationalbibliothek lists this publication in the Deutsche Nationalbibliografie; detailed bibliographic data are available in the Internet at http://dnb.d-nb.de .

D 82 (Diss. RWTH Aachen University, 2014)

ISBN 978-3-8325-4090-6
ISSN 1866-3052
Vol. 22

Logos Verlag Berlin GmbH
Comeniushof, Gubener Str. 47,
D-10243 Berlin
Tel.:   +49 (0)30 / 42 85 10 90
Fax: +49 (0)30 / 42 85 10 92
http://www.logos-verlag.de

DIRECTIVITY PATTERNS FOR ROOM ACOUSTICAL MEASUREMENTS AND SIMULATIONS

Von der Fakultät für Elektrotechnik und Informationstechnik der
Rheinischen-Westfälischen Technischen Hochschule Aachen
zur Erlangung des akademischen Grades eines

DOKTORS DER INGENIEURWISSENSCHAFTEN

genehmigte Dissertation

vorgelegt von

Diplom–Ingenieur

Martin Pollow

aus Rosenheim

Berichter:

Universitätsprofessor Dr. rer. nat. Michael Vorländer
Universitätsprofessor Dr.-Ing. Peter Vary

Tag der mündlichen Prüfung: 29. September 2014

Diese Dissertation ist auf den Internetseiten der Hochschulbibliothek online verfügbar.

meinen Eltern gewidmet

Abstract

Acoustic sources and receivers possess distinct directivity patterns that quantify their directional dependent behavior. In common room acoustical applications these directivity patterns are often neglected, although they are known to have considerable influence on the obtained results. This thesis presents methods to obtain and implement source and receiver directivity patterns for room acoustical measurements and simulations. A typical complex scenario for room acoustics consists of a concert hall that is excited by various natural sound sources (such as musical instruments) with the sound received by human listeners. Therefore, a large number of natural sound sources has been measured, analyzed and processed in order to assemble a directivity database. The sound reception of the human listeners is described by the head-related transfer functions (HRTFs) that are derived from numerical simulations or measurements of artificial heads and human individuals.

Both source and receiver directivity patterns can be represented as spherical wave spectra (SWS) using a decomposition of the angular functions into the set of orthonormal spherical harmonics. In this domain angular interpolation and range extrapolation can be implemented conveniently, yielding physically correct results for sampling schemes that are sufficiently dense in order to avoid spatial aliasing artifacts. Suitable regularization allows to derive the SWS of spatially incomplete data (with missing information at some directions) or data that suffers from measurement uncertainties.

The measurement of room impulse responses for arbitrary directivity patterns is performed using specialized spherical loudspeaker arrays to provide broadband excitation in terms of temporal and spatial frequency components. Assuming linear and time-invariant systems a sequential measurement approach greatly enhances the maximum resolution of the synthesized directivity patterns. For room acoustical simulations that include source and/or receiver directivity patterns the particle based methods can be extended by multiplication in the spatial domain. Wave based analytical simulations allow to implement arbitrary directivity patterns by the computation of the Cartesian derivatives of the room's eigenmodes. The results obtained in this thesis can be used to enhance the auralization of rooms and to analyze the perceptional impact of source and receiver directivity in room acoustical applications.

Contents

1

Introduction

Sound sources can usually be distinguished by their spectral color and other spectral or temporal features that can be obtained when listening to a mono sound recording of that source. However, the spatial variations are missing when using only a single channel: As most sound sources do not radiate omnidirectionally (i.e. uniformly in all directions) they show specific directivity patterns having a different sound color for the different directions of sound radiation.

Assume exemplarily a human speaker that is either facing the listener or looking away from the listener. For free-field conditions or an acoustic scene of low reverberation, this head orientation obviously has great influence on the perceived sound. For more reverberant environments the difference diminishes, so the audibility of the directivity patterns is closely tied to the acoustics of a room. The directional dependent behavior of the ears of a human listener can be considered as another directivity pattern, which occurs at the receiver side. Combining source and receiver directivity in measurement or simulation thus allows to obtain a binaural recording of the source in the given room.

In room acoustical measurements and simulations these directivity patterns are generally not taken into account. While simulations commonly assume omnidirectional directivity patterns, the directionality of the sensor used in the measurements contributes with its directivity to the measurement result. If performing room acoustical measurements according to the international standard ISO 3382 [Iso], sources and receivers with omnidirectional directivity patterns have to be used for the measurements in order to gain comparable results that are largely independent of the used type of the sensors and its orientation.

According to literature, directivity patterns of sources and receivers influence the room transfer function and their variation can cause audible differences: Otondo et al. [Oto04] state that the directivity patterns of musical instruments are pronounced and have "direct influence on the distribution of acoustical parameters in a room". Wang et al. [Wan08] performed listening tests using omnidirectional sound sources, octave band averaged directivity patterns (cf. also [Mey78]), and highly focused sources in rather diffuse rooms. They conclude that directivity patterns of strongly directive sources are audible, as the participants could discriminate differences in realism, reverberation and clarity. These differences contribute to the room acoustical parameters when derived from the measurements.

The goal of this thesis is to implement directivity patterns of sources and receivers in a physically correct manner for room acoustical applications. The organization of the work can be summarized as follows:

In Chapter 2 the required theory and tools are derived for the subsequent chapters. The transformation of the spatial directivity patterns into the spherical Fourier domain allows to represent a directivity as spherical wave spectrum (SWS). Compact loudspeaker array systems can be described by exterior boundary value problems to facilitate efficient analytic computation. The impact of a spatial sampling of generally continuous spherical functions is discussed. The content of this chapter comprises the theoretic background of this work and is referenced later from the following chapters.

Chapter 3 describes the methods how to obtain directivity patterns of natural sound sources using a surrounding spherical microphone array. The directivity patterns are analyzed and processed in order to provide data in the required format of room acoustical applications. The feasibility of averaging approaches depends on the principle of sound generation. At the end of the acoustical transfer path often a human listener receives the sound from all directions in a binaural audio scene.

In Chapter 4 the concepts of Fourier acoustics are applied to binaural technology, offering new approaches for common problems in the processing of head-related transfer functions (HRTFs). The proximity effect of sources close to the ear can be solved using analytical tools. Hereby, a variation of the coordinate center used as focal point for the spherical harmonic transform (SHT) seems to be beneficial for the calculations. Having obtained realistic source and receiver directivity patterns, the room impulse responses (RIRs) can be obtained with respect to the given directivity patterns using either room acoustic measurement or simulation. Both methods allow to obtain the RIR for a given combination of source and receiver directivity patterns and are described in the remaining part of the thesis.

Chapter 5 deals with the design of measurement sources with variable directivity for room acoustic applications. Using a sequential measurement approach, the measurement devices can be combined with a computerized positioning system for greatly enhanced spatial resolution. Arbitrary directivity patterns can be synthesized from the measured responses with given directivity, either by a calculation using complex reconstruction of directivity patterns or using averaged patterns without phase relations. Experimental results show that this method has great potential for the implementation of arbitrary directivity in post-processing using a set of transfer path measurements with known directivity.

Chapter 6 describes how to include arbitrary source and receiver directivity patterns into room acoustic simulation software. In wave based simulations the arbitrary patterns can be represented as physical multipoles, providing analytic solutions for simple geometries by calculating the Cartesian derivatives of the room's eigenfunctions. In particle based simulations source and receiver directivity patterns are implemented in the spatial domain, weighting the outgoing or incoming rays with the directivity value of the particular direction.

Including directivity pattern into room acoustical applications such as auralization is an active topic of research [Dal93; Beh02; Oto04; War04]. Its significance has been known for a long time, but current advances in technology of measurement techniques and simulation opened this research topic in the recent years for new applications and practical experiments.

2

Fundamentals and preliminaries

In this chapter the mathematical preliminaries are given that allow to facilitate the methods of Fourier acoustics for spatial audio processing. The theoretical concepts required for signal processing of spherical functions are discussed briefly. For a more profound theoretical overview, the reader is referred to literature listed in the references and mentioned in this chapter.

Written works considering the theory of acoustics in spherical geometry which are worth mentioning in particular include the following: Williams [Wil99] comprises a complete description of Fourier acoustics in Cartesian, cylindrical and spherical coordinates and constitutes probably one of the most complete descriptions of using spatial Fourier methods for applications in acoustics. Zotter [Zot09a] contributed a comprehensive work on sound recording and recreation using spherical arrays, including considerations on the choice of spatial sampling schemes and many other topics. Rafaely advances the research in spherical acoustics with a focus on topics regarding the use of spherical microphone arrays [Raf04; Raf05; Raf07]. Duraiswami et al. [Dur04] and Gumerov et al. [Gum05] approach this topic from a computational point of view and provide numerous recurrence relations and mathematical methods for deriving suitable formula for specific problems.

2.1. Definitions of directivity

The directional dependance of the radiation and reception of sound sources and receivers, respectively, can both be quantified by their directivity. In the literature several definitions can be found for directivity patterns, all of these definitions being generally frequency dependent functions.

Mechel [Mec08] states various definitions for expressing directivity patterns, with some of them given here briefly. All these measures have in common that the directivity pattern ought to be measured in the far-field of the source. The *directivity factor* regards a complex frequency spectrum, defined as

$$D_0(\theta, \phi) = \frac{p(r, \theta, \phi)}{p(r, \theta_0, \phi_0)} \tag{2.1}$$

with $p(r, \theta, \phi)$ being the complex sound pressure spectrum measured in the far-field on the surface of a sphere with radius r. It defines the directivity factor as the ratio between the pressure encountered in all directions and the pressure encountered in a reference direction (θ_0, ϕ_0). Mechel also gives the alternative definition of the directivity factor as

$$D_{0,\text{magn}}(\theta, \phi) = \frac{|p(r, \theta, \phi)|}{|p(r, \theta_0, \phi_0)|} \tag{2.2}$$

being the quotient of the magnitude values. Both definitions are functions of angles and frequency, with the latter being usually omitted in this work for notational ease.

These definitions of directivity are useful for the application of dry[1] sound recordings performed at a single direction. By selecting the primary recording direction as reference direction for the directivity factor, there is no colorization of the resulting sound. The directivity factor according to Mechel [Mec08] is equivalent to the monaural transfer function, one of the possible definition of the head-related transfer function (HRTF) according to Blauert [Bla97] (cf. also Eq. (4.2)). This relation shows the equivalence of binaural signal processing and the processing of directivity patterns with the methods of Fourier acoustics.

The *directivity value* is defined as the squared magnitude values at different directions in relation to their geometric average and can be stated as

$$D_m(\theta, \phi) = \frac{|p(r, \theta, \phi)|^2}{\langle |p(r, \theta', \phi')|^2 \rangle_{(\theta', \phi')}} \tag{2.3}$$

with $\langle \cdot \rangle_{(\theta', \phi')}$ being the average over all directions. The multiplication of the directivity value thus does not change the total radiated sound power of the source, but expresses the enhancement or reduction of sound radiation for all directions. Expressed as a logarithmic term the *directivity* can be given as

$$D_{Lm}(\theta, \phi) = 10 \log_{10} \frac{|p(r, \theta, \phi)|^2}{\langle |p(r, \theta', \phi')|^2 \rangle_{(\theta', \phi')}} \text{ dB}, \tag{2.4}$$

---

[1] A dry recording only contains direct sound without reverberation and without reflections of the emitted sound. For the purpose of measurements dry recordings are usually performed in an anechoic chamber.

which corresponds to the logarithmic diffuse-field equalized set of HRTFs, also called the *directional transfer functions* (DTFs) [Mid99].

2.2. Fourier transform of signals

The *Fourier transform* (FT) is one of the most commonly employed integral transforms in signal processing. It provides the representation of a (time) signal $f(t)$ in terms of its spectral components

$$\hat{f}(\omega) = \mathcal{F}\{f(t)\} = \int\limits_{-\infty}^{\infty} f(t)\, e^{-j\omega t}\, dt \tag{2.5}$$

with $\hat{f}(\omega)$ being the Fourier domain of $f(t)$ and $\omega = 2\pi f$ denoting the angular frequency. The inverse Fourier transform is given as

$$f(t) = \mathcal{F}^{-1}\left\{\hat{f}(\omega)\right\} = \frac{1}{2\pi} \int\limits_{-\infty}^{\infty} \hat{f}(\omega)\, e^{j\omega t}\, d\omega. \tag{2.6}$$

These two representations regard the time domain and the frequency domain of a signal. Assuming real-valued time signals, it is sufficient to limit the focus to the positive part of the spectrum. The values at negative frequencies can be derived from the positive frequencies by the application of the symmetry relation

$$\hat{f}(\omega) = \overline{\hat{f}(-\omega)} \tag{2.7}$$

for real-valued time signals with the overbar denoting the complex conjugate.

Having a discretely sampled signal in the time domain the *Fourier transform of discrete time signals* (FTDS) can be employed for the transform. The spectrum can be computed as a sum over the discrete time values at

$$t = \nu T \tag{2.8}$$

with ν being an integer value, defined as

$$\hat{f}(\omega) = \sum_{\nu=-\infty}^{\infty} f(\nu T)\, e^{-j\omega\nu T} \tag{2.9}$$

with the constant sampling period $T = \frac{1}{f_s}$ and f_s being the sampling rate of the equidistant discretization in time domain. The representation of the discretized signal in the Fourier domain is a continuous and periodic function, which is equal to a periodic summation of the Fourier transform of the continuous time signal.

Signals that fulfill the *Nyquist-Shannon sampling theorem* can be reconstructed without loss of information, as no spectral overlap occurs. Using the *Nyquist frequency* $f_N = \frac{f_s}{2}$ the Fourier transform of the signal has to fulfill

$$\hat{f}(\omega) = 0 \quad \text{for all} \quad |f| > f_N. \tag{2.10}$$

Applying discretization in frequency domain this yields the *discrete Fourier transform* (DFT) which is precise for signals of finite duration. The DFT and the more computationally efficient *fast Fourier transform* (FFT) are commonly found in literature and prove to be useful in many practical applications.

Using the relation

$$\omega = f_s \frac{2\pi}{M} \mu \tag{2.11}$$

with $\frac{2\pi}{M}$ being the distance between two normalized frequencies and μ being an integer value enumerating the M spectral bins, the discrete versions of the Fourier transform in Eq. (2.5) and Eq. (2.6) can be given with the integral in the equations replaced by a discrete sums.

Employing Eq. (2.11) and Eq. (2.8) the spectrum $\hat{f}(\omega)$ can be described by $\hat{f}(\mu)$ and the time domain signal $f(t)$ by $f(\nu)$, yielding the DFT as

$$\hat{f}(\mu) = \sum_{\nu=0}^{M-1} f(\nu) e^{j2\pi \frac{\mu\nu}{M}} \tag{2.12}$$

for the forward transform with $\nu = 0\ldots(M-1)$ being the number of time samples and $\mu = 0\ldots(M-1)$ being the number of spectral bins. The inverse discrete Fourier transform is given as [Var06]

$$f(\nu) = \frac{1}{M} \sum_{\mu=0}^{M-1} \hat{f}(\mu) e^{-j2\pi \frac{\mu\nu}{M}}. \tag{2.13}$$

These transforms are computed without aliasing artifacts if both time and frequency signal can be fully represented by these limited number of M time samples and M frequency bins. As the Fourier transform (as well as the DFT or FFT in case of band limited and finite signals) regards two equivalent domains of the same signal, both domains are used in this work without explicit notice of the transformation applied.

Beside the commonly used relation between a time signal and its spectral representation, the Fourier transform also allows generalization using higher dimensional functions. In the following section the application of the Fourier transform on space domain signals is discussed, in particular the spatial (two-dimensional) Fourier transform for functions defined on a spherical surface.

2.3. Fourier acoustics in spherical coordinates

The Fourier transform in the spatial domain is commonly applied on multi-dimensional functions. The analytic expressions required to perform Fourier acoustics in the space domain simplify considerably for lower dimensional cases, such as a two dimensional description of sound fields. Depending on used symmetries and the choice of the used coordinate system, the feasibility of analytic solutions vary. In this section the fundamentals of Fourier acoustics in spherical coordinates are presented.

2.3.1. Wave equation and separation of variables

The propagation of acoustic waves in a viscous medium can be described by the wave equation as

$$\Delta p = \frac{1}{c^2} \frac{\partial^2 p}{\partial t^2} \tag{2.14}$$

with the wave number $k = \frac{\omega}{c}$ and the Laplace operator $\Delta = \nabla^2$. In Cartesian coordinates the wave equation is given as the differential equation of second order as

$$\frac{\partial^2 p}{\partial x^2} + \frac{\partial^2 p}{\partial y^2} + \frac{\partial^2 p}{\partial z^2} = \frac{1}{c^2} \frac{\partial^2 p}{\partial t^2} \tag{2.15}$$

with solutions for the pressure p as a function of the three-dimensional space and time. Cartesian coordinates, e.g., are well suited for plane wave propagation as the direction of the traveling wave can easily be aligned with an axis of the coordinate system, while the other two axes span a plane of constant phase fronts for the plane wave. For spherical waves, however, the spherical coordinate system with the coordinate origin being the center point of radiation is the most convenient choice. Using the definition of the Laplace operator in spherical coordinates, the wave equation

$$\frac{1}{r^2} \frac{\partial}{\partial r} \left(r^2 \frac{\partial p}{\partial r} \right) + \frac{1}{r^2 \sin \theta} \frac{\partial}{\partial \theta} \left(\sin \theta \frac{\partial p}{\partial \theta} \right) + \frac{1}{r^2 \sin^2 \theta} \frac{\partial^2 p}{\partial \phi^2} = \frac{1}{c^2} \frac{\partial^2 p}{\partial t^2} \tag{2.16}$$

9

can be derived [Wil99].

The solutions of this equation can be found by applying the concept of separation of variables and can thus be formulated as a product of three independent variables in space plus one independent variable for the time dependence:

$$p(r, \theta, \phi, t) = R(r) \, \Theta(\theta) \, \Phi(\phi) \, T(t) \tag{2.17}$$

The separation approach yields four ordinary differential equations [Wil99]:

$$\frac{d^2 \Phi}{d\phi} + m^2 \Phi = 0 \tag{2.18}$$

$$\frac{1}{\sin \theta} \frac{d}{d\theta} \left(\sin \theta \frac{d\Theta}{d\theta} \right) + \left[n(n+1) - \frac{m^2}{\sin^2 \theta} \right] \Theta = 0 \tag{2.19}$$

$$\frac{1}{r^2} \frac{d}{dr} \left(r^2 \frac{dR}{dr} \right) + k^2 R - \frac{n(n+1)}{r^2} R = 0 \tag{2.20}$$

$$\frac{1}{c^2} \frac{d^2 T}{dt^2} + k^2 T = 0 \tag{2.21}$$

with the variables m and n being integers. Each of the four terms must satisfy the wave equation. Having one independent variable each, terms for the general solutions can be given:

$$\Phi(\phi) = \Phi_1 e^{jm\phi} + \Phi_2 e^{-jm\phi} \tag{2.22}$$

$$\Theta(\theta) = \Theta_1 P_n^m(\cos \theta) + \Theta_2 Q_n^m(\cos \theta) \tag{2.23}$$

$$R(r) = R_1 h_n^{(1)}(kr) + R_2 h_n^{(2)}(kr) \tag{2.24}$$

$$T(\omega) = T_1 e^{j\omega t} + T_2 e^{-j\omega t} \tag{2.25}$$

Hereby $P_n^m(x)$ and $Q_n^m(x)$ are the *associated Legendre functions* of first and second kind and $h_n^{(1)}(x)$ and $h_n^{(2)}(x)$ are the *spherical Hankel functions* of first and second kind. In order to obtain only functions without discontinuity in azimuthal direction, the variable m needs to be of integer value. Allowing both positive and negative values for m, only one summand is sufficient for a complete description of all possible solutions, thus Φ_2 can be set to zero. As the associated Legendre functions of second kind possess singularities at their poles at $\theta = 0$ and $\theta = \pi$ the term Θ_2 can also be set to zero. The index variable n of the Legendre functions has to be an integer value for valid solutions without singularity at the poles [Wil99].

The associated Legendre functions of first kind are defined for positive degrees m as

$$P_n^m(x) = (-1)^m (1 - x^2)^{m/2} \frac{\mathrm{d}^m}{\mathrm{d}x^m} P_n(x). \tag{2.26}$$

The functions for negative degrees can be derived by the relation

$$P_n^{-m} = (-1)^m \frac{(n-m)!}{(n+m)!} P_n^m(x) \tag{2.27}$$

with

$$P_n(x) = \frac{1}{2^n n!} \frac{\mathrm{d}^n}{\mathrm{d}x^n} (x^2 - 1)^n \tag{2.28}$$

being the orthogonal set of *Legendre polynomials* of order n [Wil99].

The spherical Hankel functions represent both incoming and outgoing waves, depending on the sign convention used for the time constant. Eq. (2.25) describes the time dependence in which the anti-causal solution can be disregarded. According to the used sign convention the constant T_2 can be set to zero in order to describe causal time signals only.[2]

The separation of variables described here is a fundamental concept in Fourier acoustics and allows to derive analytical solutions for the tempo-spatial structure of acoustic fields. The results of this section regard the base for deriving the spherical harmonics as a set of scalar orthonormal functions on the sphere (by combining the angular parts) as shown in the following section. The angular terms in combination with the radial part can be considered as *multipole expansion* and will be described in Sec. 2.3.4.

2.3.2. Spherical harmonics

Both Eq. (2.22) and Eq. (2.23) are functions with the argument of an angle. They can be combined multiplicatively (using $\Phi_2 = \Theta_2 = 0$) to a set of functions called *spherical harmonics* (SH) of order n and degree m, defined as

$$Y_n^m(\theta, \phi) = \sqrt{\frac{(2n+1)}{4\pi} \frac{(n-m)!}{(n+m)!}} \cdot P_n^m(\cos\theta) \cdot \mathrm{e}^{\mathrm{j}m\phi} \tag{2.29}$$

The square root term in the definition of the spherical harmonics is the chosen product of Φ_1 and Θ_1 and provides for the orthonormality of the spherical har-

---

[2]This sign convention is the commonly used phasor in engineering and is implemented in the measurement and signal processing software packages, Monkey Forest [Mül99] and the ITA-Toolbox for Matlab [Die12a].

monics.[3] The spherical harmonics functions of negative order can be computed from the functions of positive order using the relation

$$Y_n^m(\theta, \phi) = (-1)^m \cdot \overline{Y_n^{-m}(\theta, \phi)} \tag{2.30}$$

with the overbar denoting the complex conjugate. It can be seen that the absolute value of the functions is not changed by this conversion rule, so that a function of negative degree differs from the function of identical order and symmetric positive degree only by its phase. In Fig. 2.1 the set of the spherical harmonics up to an order of two are depicted. The color is used to describe the phase of the function, while the radius quantifies the magnitude of the function at the given direction.

A fundamental property of the spherical harmonics functions is their *orthonormality*. The inner product of two spherical harmonics equals unity for identical order and degree and zero for differing spherical harmonics. This can be expressed mathematically as

$$\oint_{S^2} Y_n^m(\theta, \phi) \overline{Y_{n'}^{m'}(\theta, \phi)} \, \mathrm{d}\Omega = \langle Y_n^m \mid Y_{n'}^{m'} \rangle = \delta_{nn'} \delta_{mm'} \tag{2.31}$$

with

$$\oint_{S^2} \mathrm{d}\Omega = \int_0^{2\pi} \int_0^{\pi} \sin\theta \, \mathrm{d}\theta \, \mathrm{d}\phi \tag{2.32}$$

being the integral over the surface S^2 of the unit sphere and the *Kronecker delta function* defined as

$$\delta_{xy} = \begin{cases} 1 & \text{for} \quad x = y \\ 0 & \text{for} \quad x \neq y \end{cases} . \tag{2.33}$$

According to Williams [Wil99] the set of spherical harmonics constitute a vector basis for square-integrable functions. In order to fulfill that requirement the integral of the squared magnitudes of the function has to be bound to a limited value:

$$\oint_{S^2} |f(\theta, \phi)|^2 \, \mathrm{d}\Omega < \infty \tag{2.34}$$

---

[3] Other definitions of the spherical harmonics exist, either with different sign convention or with a different normalization. In the latter case the spherical harmonics regard an orthogonal basis that is not orthonormal.

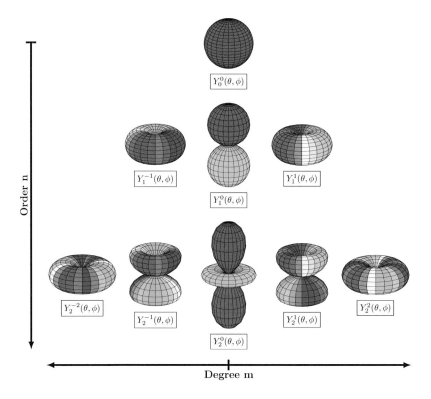

Figure 2.1.: Complex spherical harmonics as balloon plots; radius denotes the magnitude and color denotes the phase for a given direction.

2.3.3. Spherical wave spectrum

Williams [Wil99] further defines the *spherical wave spectrum* (SWS) f_{nm} as the spherical harmonic transform of an arbitrary square-integrable spherical function $f(\theta, \phi)$. The spherical harmonics are used as basis functions for the decomposition into a set of SH coefficients. A spherical scalar function can thus be represented in two equivalent domains: in the *SH domain* or in the *spatial domain*, analogous to time and frequency domain when using the Fourier transform. Here the pair of transformations for spherical functions and a relation between the two domains is given.

Spherical harmonic transform

The *spherical harmonic transform* (SHT) is defined as

$$f_{nm} = \mathcal{S}\left\{f(\theta, \phi)\right\} = \oint_{S^2} f(\theta, \phi) \cdot \overline{Y_n^m(\theta, \phi)} \, d\Omega. \tag{2.35}$$

This operation is generally precise if the integral can be solved in an exact way. In practice we are usually dealing with spatially sampled directivity patterns, so that the integral has to be approximated by a summation over a set of discrete points. The issues that possibly arise (such as e.g. aliasing artifacts) are addressed in Sec. 2.5.

Inverse spherical harmonic transform

The *inverse spherical harmonic transform* (ISHT) is given as

$$f(\theta, \phi) = \mathcal{S}^{-1}\left\{f_{nm}\right\} = \sum_{n=0}^{\infty} \sum_{m=-n}^{n} f_{nm} \cdot Y_n^m(\theta, \phi). \tag{2.36}$$

This operation is precise if the outer summation is performed up to an order of infinity. For order limited SWS, the infinite sum can be replaced by a finite sum up the maximum order of the SWS.

Parseval's identity

As in the one-dimensional case Parseval's identity is valid for the Fourier transform on the sphere [Bro00]:

$$\oint_{S^2} |f(\theta, \phi)|^2 \, d\Omega = \sum_{n=0}^{\infty} \sum_{m=-n}^{n} |f_{nm}|^2 \tag{2.37}$$

This equation shows that the squared magnitudes of a spherical function integrated over the complete solid angle equals the squared magnitudes of its spherical wave spectrum.[4]

---

[4] As the squared magnitude values are proportional to the energy values of a signal, the term "signal energy" is used even if the required normalization with an impedance for conversion to physical energy is being done.

2.3.4. Boundary value problems

Boundary value problems in acoustics consist of a sound field description using the Helmholtz equation and a set of known values given at a boundary surface. If the boundary surface can be described in an appropriate coordinate system by a two dimensional subset of the three dimensional space, efficient analytic solutions for the determination of the complete sound field exist. For spherically shaped boundary surfaces, spherical coordinates are the natural choice and deliver such fast and precise analytic solutions.

Three cases of boundary value problems can be differentiated as follows:[5]

Exterior problems

Exterior problems occur when all sources are confined within a certain area in space. In Fig. 2.2a the sources are confined within the gray spherical area of radius a. The area outside of this sphere is considered source free and the problem satisfies the Sommerfeld radiation condition (no incoming radiation from infinity).

If the pressure field (Neumann boundary condition) or the surface velocity (Dirichlet boundary condition) is known on a separable surface, the complete field outside of $r > a$ can be determined, an area marked in Fig. 2.2a as region of validity, cf. [Zot09a]. This kind of problem occurs e.g. when calculating the radiation of a sound source at different radial distances from data obtained at a single distance. Using reciprocity the solution of the exterior problem can also be applied to sound receivers with complex directivity patterns.

The sound pressure field can be described by the *multipole expansion* of the field into a set of *singular solutions* of the Helmholtz equation as

$$p(r, \theta, \phi) = \sum_{n=0}^{\infty} \sum_{m=-n}^{n} c_{nm} h_n(kr) Y_n^m(\theta, \phi) \qquad (2.38)$$

with the expansion coefficients c_{nm} being generally functions of frequency and k being the wave number. This multipole expansion combines the spherical wave spectrum with a term for radial wave propagation. Exploiting the orthonormality of the spherical harmonics these coefficients can be derived by the multiplication

---

[5]As of reciprocity these problems can be formulated for sources and receivers, here the problems for sources are formulated.

of the conjugated complex of the spherical harmonics and an integration over the 2-sphere, resulting in the following relation [Wil99]:

$$c_{nm} = \frac{1}{h_n(kr)} \oint_{S^2} p(r, \theta, \phi) Y_n^m(\theta, \phi) \, d\Omega \qquad (2.39)$$

In these equations the spherical Hankel functions for an outgoing wave have to be employed. With the used sign conventions these are the spherical Hankel function of 2nd kind, abbreviated as $h_n(\cdot) = h_n^2(\cdot)$.

For a given spherical wave spectrum obtained at a radial distance $r_0 > a$ from the focal point in the center of the coordinate system an extrapolated version for the radial distance $r_1 > a$ can be calculated. This calculation is valid in the white (source free) area in Fig. 2.2a and is performed in the spherical harmonic domain. The spherical wave spectrum of the extrapolated function is multiplied with the fraction of the spherical Hankel functions of the two radii as

$$c_{nm}(r_1, k) = c_{nm}(r_0, k) \frac{h_n(kr_1)}{h_n(kr_0)}. \qquad (2.40)$$

The fraction of Hankel functions contains the $1/r$ decay, as well as potential near-field components of the higher orders of the SWS.

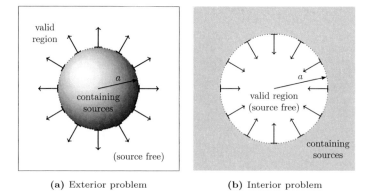

(a) Exterior problem   (b) Interior problem

Figure 2.2.: Boundary value problems with the gray area containing all sources and the white area being considered source free.

Interior problems

An interior problem (cf. Fig. 2.2b) occurs for a source-free region surrounded by acoustical sources. Also in this case, the knowledge of the sound pressure (Neumann boundary condition) or the particle velocity (Dirichlet boundary condition) on a separable surface is sufficient to determine the sound field in the source-free area within a confined space between the sources.

The sound pressure field inside the confined area can be described by the series of

$$p(r, \theta, \phi) = \sum_{n=0}^{\infty} \sum_{m=-n}^{n} b_{nm} j_n(kr) Y_n^m(\theta, \phi) \tag{2.41}$$

with the expansion coefficients b_{nm}. In comparison to the exterior problem with the singular solution given by the expansion coefficients c_{nm}, the spherical Hankel functions h_n are replaced by the spherical Bessel functions j_n, yielding the *regular solutions* of the Helmholtz equation given by b_{nm}. As the name suggests this solution does not possess a singularity at the origin of the coordinate system, yielding finite values for all points in the region of validity.

The derived coefficients are functions of the frequency and can be found analogously to the exterior problem as [Wil99]:

$$b_{nm} = \frac{1}{j_n(kr)} \oint_{S^2} p(r, \theta, \phi) Y_n^m(\theta, \phi) \, d\Omega \tag{2.42}$$

Analogously to the exterior problem, the radial distance can be extrapolated in the spherical harmonics domain as

$$b_{nm}(r_1, k) = b_{nm}(r_0, k) \frac{j_n(kr_1)}{j_n(kr_0)} \tag{2.43}$$

with the fraction of spherical Bessel functions giving a frequency dependent and order dependent scaling factor.

Mixed problems

Mixed problems can be regarded as a combination of exterior and interior problems. In case of spherical coordinates they provide solutions e.g. for a spherical shell (three-dimensional annulus) that either contains all sources within or outside of its volume. For details confer to literature [Wil99; Gum03; Zot09a].

2.3.5. Transformation of acoustic spherical fields

Acoustic fields can be expressed by their Fourier expansion using the coefficients c_{nm} or b_{nm} as defined by Eq. (2.39) and Eq. (2.42). Analytical expressions exist to perform translation and rotation in the Fourier domain or in the spatial domain. While the rotation of spherical wave spectra or multipole expansions are computationally straight-forward, the translation in the SH domain can be computationally very demanding. In this section a short overview of the methods for source translation and rotation is given.

Translation

In order to translate the acoustic field from one focal point to another, its multipole expansion coefficients can be transformed using recurrence relations as given e.g. in Gumerov et al. [Gum03] and Zotter [Zot09a]. The translation can be performed on the acoustic field represented as a weighted set of singular and regular solutions of the Helmholtz equation. This can be done for the translation in any direction using recurrence relations that demand a comparatively high computational effort for high orders. While the number of coefficients rise quadratically with rising order, the number of required operations for a generic translation rises with the power of five [Gum03].[6]

A significant speedup can be achieved by applying prior rotation so that the direction of translation is aligned with the z-axis of the Cartesian coordinate system. A translation in that direction is computationally much more efficient than arbitrary translation [Zot09a; Kle12a]. After the rotation and the application of the translation exploiting the symmetry to the z-axis, the inverse rotation restores the orientation of the original problem.

For fast computation the translation can also be approximated in the spatial domain, although it is generally not possible to take near-field effects into account. Using a far-field approximation of a singular source translated from the distance r_0 to r_1 all higher orders are disregarded, yielding

$$p(r_1) = p(r_0) \cdot \frac{h_0(kr_1)}{h_0(kr_0)} = p(r_0) \cdot \frac{r_0}{r_1} e^{jk(r_0 - r_1)}. \qquad (2.44)$$

---

[6]Calculation times for generic translations for higher orders generally range from several hours to several days of calculation time on a standard laptop.

Calculating source translation in the SH domain, the order dependent effects are taken into account. A spherical function that has a limited rate of change over the spatial angle typically has an order limited spherical wave spectrum (SWS). After applying the translation the SWS is modified for the differing focal point. While functions such as the sound pressure created from a compact sound source are generally limited in order (cf. Sec. 2.3.8), a translation can lead to a displaced source where a non-optimal focal point is used for the transformations, yielding much higher orders of the SWS.

Rotation

In the spatial domain any possible rotation can be expressed by three single rotations about the Cartesian coordinate axes. These rotations can be quantified by the Euler angles α, β and γ, denoting rotations around the z-axis, the y-axis and a second time around the z-axis, respectively.

Any point in space can be rotated by the multiplication of its Cartesian coordinates with a 3×3 matrix with a determinant of unity that can be composed by the multiplication of three rotation matrixes $\mathbf{R_z}(\alpha)$, $\mathbf{R_y}(\beta)$ and $\mathbf{R_z}(\gamma)$ defined as [Kos03]

$$\mathbf{R_z}(\alpha) = \begin{pmatrix} \cos\alpha & -\sin\alpha & 0 \\ \sin\alpha & \cos\alpha & 0 \\ 0 & 0 & 1 \end{pmatrix} \qquad \text{and} \qquad (2.45)$$

$$\mathbf{R_y}(\beta) = \begin{pmatrix} \cos\beta & 0 & \sin\beta \\ 0 & 1 & 0 \\ -\sin\beta & 0 & \cos\beta \end{pmatrix}. \qquad (2.46)$$

A point in space defined by its Cartesian position vector

$$\mathbf{r} = x\,\vec{e}_x + y\,\vec{e}_y + z\,\vec{e}_z \qquad (2.47)$$

can be transformed to the rotated position by a subsequent multiplication with the rotation matrices as

$$\mathbf{r}_{\text{rot}} = \mathbf{R_z}(\alpha) \cdot \mathbf{R_y}(\beta) \cdot \mathbf{R_z}(\gamma) \cdot \mathbf{r}. \qquad (2.48)$$

Alternatively quaternions can be used for the description of arbitrary rotations in the spatial domain [Kui99].

Rotation can also be implemented in the Fourier domain. As any set of spherical harmonics of a given order forms an invariant subspace, the signal energy of a specific order of a spherical function represented by its spherical wave spectrum does not change by rotation. The distribution of energy in the various degrees of a fixed order varies, but the total energy is constant.

The rotation is applied to a given spherical wave spectrum or multipole expansion by the Wigner-D function defined for the given rotation expressed by the Euler angles (α, β, γ) as

$$D_{k,m}^{(n)}(\alpha, \beta, \gamma) = e^{-jk\alpha} \cdot d_{k,m}^{(n)}(\beta) \cdot e^{-jm\gamma}. \tag{2.49}$$

with n, m and k being integer values and $d_{k,m}^{(n)}(\beta)$ being the reduced (real-valued) Wigner-d function, which can be computed by a set of recurrence formulas and explicit starting values as given in Kostelec et al. [Kos03] or Pendleton [Pen03]. The rotation can then be applied in the spherical harmonic domain as

$$f_{nm,\text{rot}} = \sum_{k=-n}^{n} f_{nk} \cdot D_{k,m}^{(n)}(\alpha, \beta, \gamma). \tag{2.50}$$

As the squared absolute values of the Wigner-D function add up for all values of k to unity, the total signal energy stored in a certain SH order of the spherical wave spectrum is not changed by the rotation:

$$\sum_{k=-n}^{n} \left| D_{k,m}^{(n)} \right|^2 = 1 \tag{2.51}$$

Fast algorithms for the computation of the Wigner-D can be found e.g. in Zotter [Zot09a] and Klein [Kle12a].

2.3.6. Dirac impulse on unit sphere

A useful auxiliary function in signal processing is the generalized function of the Dirac delta impulse. It can be used to sample a continuous function or to apply a time shift, exploiting its sifting property. Integrating over the Dirac delta impulse yields a value of unity. On the sphere, a Dirac impulse pointing in the direction of (θ', ϕ') can be defined as [Wil99]:

$$\delta^{(\theta', \phi')}(\theta, \phi) = \delta(\cos\theta - \cos\theta')\,\delta(\phi - \phi') \tag{2.52}$$

Williams [Wil99] also gives an alternative representation using the spherical harmonic basis functions:

$$\delta^{(\theta',\phi')}(\theta,\phi) = \sum_{n=0}^{\infty} \sum_{m=-n}^{n} \overline{Y_n^m(\theta',\phi')} Y_n^m(\theta,\phi) \tag{2.53}$$

Comparing this equation with the ISHT given in Eq. (2.36) shows the SWS of the Dirac impulse on the unit sphere as

$$\delta_{nm}^{(\theta',\phi')} = \overline{Y_n^m(\theta',\phi')}. \tag{2.54}$$

The spatial representation of the Dirac impulse on the 2-sphere integrates to unity:

$$\oint_{S^2} \delta^{(\theta',\phi')}(\theta,\phi)\,\mathrm{d}\Omega = 1 \tag{2.55}$$

This property of the S^2-Dirac impulse is analog to the one-dimensional Dirac impulse in the time domain, which also integrates to unity. It can thus also be used for sampling a function at a certain point. As in the one-dimensional case, this is done by an element-wise multiplication of the Dirac impulse with the function and by integrating over the result. A spherical function multiplied by a Dirac impulse and integrated over the sphere yields the function value at the direction of the Dirac impulse:

$$f(\theta',\phi') = \oint_{S^2} f(\theta,\phi) \cdot \delta^{(\theta',\phi')}(\theta,\phi)\,\mathrm{d}\Omega \tag{2.56}$$

2.3.7. Cross-correlation of spherical functions

In signal processing the cross-correlation is commonly used to quantify the similarity between two signals. For continuous signals the cross-correlation function of two time signals $f_1(t)$ and $f_2(t)$ is defined as a function of the time lag τ [Ohm06]

$$C(\tau) = \int_{-\infty}^{\infty} \overline{f_1(t)} f_2(\tau - t)\,\mathrm{d}t \tag{2.57}$$

with the overbar denoting the conjugated complex. The cross-correlation is identical to the inner product of two functions in the used vector space. Analogously, the inner product of two spherical functions can be regarded as their cross-correlation coefficient. It yields a measure of similarity for spherical functions and is defined as [Kos03]:

$$C(f,g) = \langle f \mid g \rangle = \oint_{S^2} \overline{f(\theta,\phi)}\, g(\theta,\phi)\, \mathrm{d}\Omega \qquad (2.58)$$

This value depends on the magnitudes of the functions $f(\theta,\phi)$ and $g(\theta,\phi)$ and changes with varying the total gain of one of these functions. In order to use it as a similarity measure for directivity patterns (which are normalized as given in the definitions in Sec. 2.1) the normalized correlation coefficient can be used.

The spherical cross-correlation coefficient can be divided by the square root of the product of the two function's squared magnitudes, yielding the normalized cross-correlation coefficient as [Pol09c]

$$\widetilde{C}(f,g) = \frac{C(f,g)}{\sqrt{E_f E_g}} \qquad (2.59)$$

with

$$E_f = \langle f \mid f \rangle = C(f,f) \qquad (2.60)$$

being the inner product of the function with itself (or the auto-correlation of the function). Using the generalization of Parseval's theorem as defined in Eq. (2.37), the spherical cross-correlation coefficient can also be defined in the spherical harmonics domain as

$$C(f,g) = \sum_{n=0}^{\infty} \sum_{m=-n}^{n} \overline{f_{nm}} g_{nm}. \qquad (2.61)$$

The normalized correlation allows convenient comparison of spherical functions, even if the data differs in sensitivity (e.g. when recorded in different uncalibrated measurement sessions) as a difference in the absolute gain does not influence the result.

2.3.8. Frequency dependent order truncation

Numerical calculations in the spherical harmonics domain are usually performed by truncating the infinite series of the SWS at a specific order. In the near-field of a source or receiver the high order components of the SWS of the directivity are highly attenuated at low frequencies and outward traveling direction, so a frequency dependent upper order truncation is suitable.

For lower maximum orders the computational cost is reduced with the price of lower accuracy. Gumerov et al. [Gum02a] describe a general convergence to the precise result for rising maximum orders, while regarding the choice of truncation number an important question for the optimal trade-off between computation speed and accuracy. Duraiswami et al. [Dur04] propose the kr-limit as rule-of-thumb for the range extrapolation of HRTFs, using only spherical harmonics up to the maximum order

$$n_{\mathrm{max}} = \lfloor kr_{\mathrm{min}} \rfloor \tag{2.62}$$

with r_{min} being the radius encompassing all sources and k being the wave number. For a specific focal point used for the expansion as spherical wave spectrum r_{min} equals to the outermost source contributing to the radiation and regards the minimum distance for the region of validity as depicted in Fig. 2.2.[7]

In literature also higher limits are found, resulting in a higher convergence to the actual result with the risk of overfitting [Dur04]. Good results have been reported by Müller-Trapet et al. [MT11] using the suggested limit raised to be two orders higher (independent of frequency).

2.4. Discrete Fourier acoustics in spherical coordinates

The theory presented until here is subject to spatially continuous calculation. Acoustic sensors usually sample the acoustic pressure field at discrete points in space. Instead of continuous functions thus often a spatially sampled subset of the values in the three dimensional space is available for further processing. As Eq. (2.35) contains a continuous integral, this operation has to be modified for discretely sampled data in order to numerically evaluate the integral in terms of a finite sum of samples. Using weights α_i for each of the I sampling points the continuous integral over S^2 shall be equally represented as discrete sum for specific sampling schemes (cf. Sec. 2.5) as

$$\oint_{S^2} f(\theta, \phi)\, \mathrm{d}\Omega \overset{!}{=} \sum_{i=1}^{I} \alpha_i f(\theta_i, \phi_i) \tag{2.63}$$

with $f_{nm} = \mathcal{S}\{f(\theta, \phi)\}$ being the spherical wave spectrum which is order limited to n_{max}, expressed as

$$f_{nm} = 0 \qquad \text{for } n > n_{\mathrm{max}}. \tag{2.64}$$

---

[7]The described order truncation is applied in Sec. 4.3 and (with a higher limit) in Sec. 5.4.3. Suitable order truncation can also be employed for computing the SHT for incomplete datasets (cf. Sec. 4.4). A weaker order limit such as the order dependent Tikhonov regularization (cf. Sec. 2.4.4) might be preferable in this case.

The weights α_i for the I sampling points can be determined by solving Eq. (2.63). For details cf. to Sec. 2.5 where spherical sampling schemes using exact quadrature are presented that yield precise forward and inverse SHT for order limited functions. Assuming such precise sampling schemes, the transforms, Parseval's theorem and orthonormality condition and can be formulated with minor modifications compared to the continuous cases in Eq. (2.31) and Eq. (2.35) to Eq. (2.37).

The *discrete spherical harmonic transform* (DSHT) is defined as

$$f_{nm} = \sum_{i=1}^{I} \alpha_i f(\theta_i, \phi_i) \overline{Y_n^m(\theta_i, \phi_i)}, \tag{2.65}$$

while the *inverse spherical harmonic transform* (ISHT) can be defined for discrete angles as

$$f(\theta_i, \phi_i) = \sum_{n=0}^{n_{\max}} \sum_{m=-n}^{n} f_{nm} \cdot Y_n^m(\theta_i, \phi_i). \tag{2.66}$$

Parseval's identity can be generalized as

$$\sum_{i=1}^{I} \alpha_i |f(\theta_i, \phi_i)|^2 = \sum_{n=0}^{n_{\max}} \sum_{m=-n}^{n} |f_{nm}|^2, \tag{2.67}$$

while the *orthonormality* is also preserved for all $n' \leq n_{\max}$ and $n \leq n_{\max}$ as

$$\sum_{i=1}^{I} \alpha_i Y_{n'}^{m'}(\theta_i, \phi_i) \overline{Y_n^m(\theta_i, \phi_i)} = \delta_{n-n'} \delta_{m-m'}. \tag{2.68}$$

2.4.1. Matrix formulation

The spatially discretized spherical harmonic basis functions can be represented in matrix notation as

$$\mathbf{Y} = \mathbf{Y}_N = \begin{pmatrix} Y_0^0(\theta_1, \phi_1) & Y_1^{-1}(\theta_1, \phi_1) & \cdots & Y_N^N(\theta_1, \phi_1) \\ Y_0^0(\theta_2, \phi_2) & Y_1^{-1}(\theta_2, \phi_2) & \cdots & Y_N^N(\theta_2, \phi_2) \\ Y_0^0(\theta_3, \phi_3) & Y_1^{-1}(\theta_3, \phi_3) & \cdots & Y_N^N(\theta_3, \phi_3) \\ \vdots & \vdots & \ddots & \vdots \\ Y_0^0(\theta_I, \phi_I) & Y_1^{-1}(\theta_I, \phi_I) & \cdots & Y_N^N(\theta_I, \phi_I) \end{pmatrix} \tag{2.69}$$

with I being the number of points used and $N = n_{\max}$ the maximum order of spherical harmonics. The linear index q is used to enumerate the columns of the matrix by order and degree of the spherical harmonics as defined (amongst others) by Zotter [Zot09a] as

$$q = n^2 + n + m + 1. \tag{2.70}$$

This linear indexing is equivalent to sort the spherical harmonics as depicted in Fig. 2.1 in reading direction from the top of the triangle line-by-line from left to right. Using this linear index the functions of second order ($n = 2$, $|m| \leq n$), i.e., the functions depicted on the third row in Fig. 2.1, have the indices $q = 2^2 + 2 + m + 1$, thus ranging from 5 to 9.

To use the order limited coefficients of the SWS by means of linear algebra, the operator $\mathrm{vec_{SH}}\{\cdot\}$ can be introduced for the conversion into a vector of coefficients as

$$\hat{\mathbf{f}} = \mathrm{vec_{SH}}\{f_{nm}\}. \tag{2.71}$$

Likewise, the operator $\mathrm{vec}\{\cdot\}$ is used to describe a function of a set of spatial sampling points as the vector

$$\mathbf{f} = \mathrm{vec}\{f\}. \tag{2.72}$$

For discretized spatial data the transformation into the order limited spherical wave spectrum regards a purely discrete transform that can be represented using matrix notation and related tools from linear algebra. The ISHT for the continuous case as defined in Eq. (2.36) can be represented as matrix multiplication with the transformation matrix $\mathbf{Y} = \mathbf{Y}_N$ containing the spatially sampled spherical harmonics as column vectors, as[8]

$$\mathbf{f} = \mathbf{Y} \cdot \hat{\mathbf{f}}. \tag{2.73}$$

This relation is valid for order limited spherical wave coefficients with $f_{nm} = 0$ for $n > N$. The SHT can be formulated as a matrix inversion of $\mathbf{Y}$ as

$$\hat{\mathbf{f}} = "\mathbf{Y}^{-1}" \cdot \mathbf{f}. \tag{2.74}$$

In general the spherical harmonics matrix $\mathbf{Y}$ is not square and thus not an invertible singular matrix. Other ways have to be found to obtain a (possibly ap-

---

[8]The matrix $\mathbf{Y}$ is chosen here to directly reflect a transformation matrix commonly used in this work. A multiplication with the matrix denotes the ISHT, while the inverse operation regards the SHT. Note that the matrix $\mathbf{Y}$ is sometimes defined differently in literature.

proximate) solution of the inverse problem. With a quadrature sampling scheme the order limited SHT can be calculated in an exact manner. The spherical wave coefficients

$$\hat{\mathbf{f}} = \mathbf{Y}^H \operatorname{diag}\{\alpha_i\} \cdot \mathbf{f}. \tag{2.75}$$

are found by an multiplication with the conjugate transpose of $\mathbf{Y}$, with a weight vector for all sampling points given as diagonal matrix and with the vector of the spatially sampled function values.

The matrix

$$\mathbf{O} = \mathbf{Y}^H \operatorname{diag}\{\alpha_i\} \mathbf{Y} \tag{2.76}$$

thus represents subsequent inverse and forward SHT. With a limited maximum order N the SHT and ISHT are precise for quadrature sampling schemes, resulting in $\mathbf{O}$ being the Identity matrix. Another possibility for computing the SHT is the use of a generalized inverse as suggested by Penrose [Pen55] (and earlier by Eliakim H. Moore), calculating a least-squares approximation for a overdetermined system of equations represented by a given matrix, for underdetermined systems the generalized inverse (also called *Moore-Penrose-Inverse*) finds the solution of minimum norm.

If a spherical wave spectrum of higher orders than the maximum order the given spherical sampling scheme can resolve is used for the SHT, aliasing artifacts occur as is discussed in Sec. 2.5

2.4.2. Error measure

The spherical residual function $r'(\theta, \phi)$ of two continuous spherical functions can be defined as

$$r'(\theta, \phi) = f(\theta, \phi) - g(\theta, \phi). \tag{2.77}$$

In case of discrete sampling of a pressure function the residual sound pressure can be formulated as a vector $\mathbf{r}_\mathrm{p} = \mathrm{vec}\{r'\}$ containing the discrete residual values for each sampling point as

$$\mathbf{r}_\mathrm{p} = \mathbf{p} - \mathbf{p}_\mathrm{orig}. \tag{2.78}$$

with $\mathbf{p}_\mathrm{orig}$ being the sound pressure free of errors and $\mathbf{p}$ being the defective sound pressure.

It can be useful to quantify the residual on a logarithmic scale in order to account for the logarithmic perception of the human hearing as used e.g. in Sec. 5.4.1.

2.4.3. Condition number

The condition number κ gives an upper limit for the amplification of errors that occur using a given function or matrix transform, thus expressing the sensitivity of the output to a change in the input. Assuming the general linear equation $\mathbf{Ax} = \mathbf{b}$ with given $\mathbf{b}$ and unknown $\mathbf{x}$, the relative error of the solution vector $\mathbf{x}$ with respect to the problem vector $\mathbf{b}$ can be expressed by the inequality [Str03]

$$\frac{||\Delta\mathbf{x}||}{||\mathbf{x}||} \leq \kappa \frac{||\Delta\mathbf{b}||}{||\mathbf{b}||} \tag{2.79}$$

with $||.||$ being the 2-norm of the vectors.

In case of a defective matrix instead of a defective input vector the error bounds can be stated as [Str03]

$$\frac{||\Delta\mathbf{x}||}{||\mathbf{x} + \Delta\mathbf{x}||} \leq \kappa \frac{||\Delta\mathbf{A}||}{||\mathbf{A}||} \tag{2.80}$$

with $||\mathbf{A}||$ being the 2-norm (and thus the largest singular value) of matrix $\mathbf{A}$. For square matrices the condition number can be defined as the norm of the matrix multiplied by the norm of its inverse [Gol96]

$$\kappa = ||\mathbf{A}|| \cdot ||\mathbf{A}^{-1}|| \tag{2.81}$$

while for non-square matrixes the inverse $\mathbf{A}^{-1}$ can be replaced with the generalized matrix inverse $\mathbf{A}^{+}$. The condition number can also be calculated by the fraction of the largest singular value of a matrix divided by the smallest singular value.

The condition number of the matrix $\mathbf{Y}_N$ can be used to derive e.g. optimal order truncation for given arbitrary geometries. In case of spatial undersampling the condition number of the matrix $\mathbf{Y}_N$ increases dramatically, indicating the ill-posedness of the SHT for a given maximum order $N = n_{\mathrm{max}}$.

2.4.4. Matrix inversion using regularization

In order to overcome the problems associated with bad conditioning, regularization methods are commonly employed for achieving better results when inverting a system of equations. This is done by using additional knowledge in order to gain sensible results for the particular physical problem that needs to be solved.

27

Using the Moore-Penrose pseudoinverse a *least mean squares* (LMS) solution with the minimum 2-norm of the residual vector is found. The generalized Tikhonov regularization regards a trade-off between minimum 2-norm of the residual vector and the 2-norm of the solution vector, with the possibility to emphasize one or the other norm [Nai97].

Extending this concept to the physical expectation of a certain spherical wave represented by their SH coefficients, allows to derive better results for the ISHT of possibly imprecise, defective data, that are possibly not even defined on a regular distribution on the surface of a sphere. This concept allows to derive solutions that minimizes the signal energy of high orders [Dur04].[9]

In acoustic applications regularization is not only used for calculating the ISHT of imprecise spatial data, but generally when matrix inversion is performed. Masiero [Mas12] employs a regularization approach in order to calculate stable filters used for the *cross-talk cancelation* (CTC) technique.

Written in matrix formulation the solution $\mathbf{x}$ for a given inverse problem $\mathbf{A}\mathbf{x} = \mathbf{b}$ can be calculated using the Moore-Penrose pseudoinverse as

$$\mathbf{x} = \mathbf{A}^{+}\mathbf{b}. \tag{2.82}$$

For a solution applying Tikhonov regularization the inverse problem can be solved by the equation

$$\mathbf{x} = \left(\mathbf{A}^{H}\mathbf{A} + \varepsilon\mathbf{D}\right)^{-1}\mathbf{A}^{H}\mathbf{b} \tag{2.83}$$

with $\varepsilon > 0$ being the regularization parameter and $\mathbf{D}$ being a diagonal matrix, often set to the Identity matrix [BI03] as

$$\mathbf{D} = \mathbf{I}. \tag{2.84}$$

Duraiswami et al. [Dur04] suggest to use an order dependent diagonal matrix for the Tikhonov regularization, defined as

$$\mathbf{D} = (1 + n(1 + n))\mathbf{I} \tag{2.85}$$

with n being the current order of the SWS. This allows to penalize higher orders more than lower orders. In combination with a frequency dependent truncation

---

[9]Examples for the effect of regularization on directivity control can be found in Sec. 4.3.3 and Sec. 5.4.1. An example for incomplete data on the sphere is given in Sec. 4.4.

of the calculation, this proves to be very effective for calculations of the SHT with imprecise data and/or incomplete spatial sampling schemes [Hea03; Dur04; MT11; Pol12a; Bro13].

2.5. Spherical sampling schemes and aliasing analysis

Directivity patterns are commonly measured on a spherical shell around a specific point that is used as focal point for the transformations of the spherical functions. In this section a selection of sampling schemes on a sphere are described that have been used for this study.

The equiangular quadrature and the Gaussian quadrature sampling scheme offer precise SHT for order limited spherical functions. Both show redundancy having double (for Gaussian) or quadruple (for equiangular) the number of points as there are coefficients in the SH domain.

The sampling scheme using hyperinterpolation on the sphere offers precise results for order limited functions without redundancy, having as many sampling points as there are coefficients in the SWS. The disadvantage of this kind of sampling is that no redundancy also means higher errors for deviations or excitation of higher orders above the order limit.

The equiangular sampling scheme using a regular angular spacing starting at the poles of the spherical coordinate system is not an exact sampling scheme. As it is often found in existing measurement data it is also analyzed in this section.

In Fig. 2.3 to Fig. 2.7 the aliasing components for these reviewed sampling schemes are depicted as greyscale levels, showing the error that occurs performing the SHT using this geometry. Using a sampling scheme that is optimized for the maximum order $N_1 = 10$, the error that occurs by (under)sampling continuous functions with a SWS of up to an order $N_2 = 20$ is depicted. The deviation from the ideal representation (perfect mapping up to order N_1 and no contributions of higher orders) yields a $N_1 \times N_2$ matrix, representing the deviations from orthonormality up to the order N_1 and aliasing artifacts for higher orders.

2.5.1. Equiangular quadrature sampling scheme

Driscoll et al. [Dri94] define a set of points with equal angular spacing in both azimuth and elevation that is suitable for numerically precise integration of order limited functions on the sphere. It consists of a rectangular sampling with $(2n_{\max} + 1)$ samples in the direction of both elevation and azimuth and provides exact weighted quadrature on the unit sphere. As all sampling points with constant elevation have identical weights, Eq. (2.63) can be reformulated as:

$$\oint_{S^2} f(\theta, \phi)\, d\Omega = \sum_{l=0}^{L} \sum_{k=0}^{K} \alpha_l f(\theta_l, \phi_k) \tag{2.86}$$

with

$$L = K = 2n_{\max} + 1 \tag{2.87}$$

and $\mathcal{S}\{f(\theta, \phi)\}$ being bound to a maximum order $n_{\max}$.

The sampling efficiency as defined by Zotter [Zot09b] for this kind of sampling is 25%. As an advantage the influence of aliasing is small compared to other types of sampling schemes [Raf07]. The aliasing components occurring for order $n > n_{\max}$ can be seen in Fig. 2.3. The Identity matrix is subtracted from the orthonormality matrix

$$\mathbf{O} = \mathbf{Y}_{N_1}^{H}\, \mathrm{diag}\{\alpha_l\}\, \mathbf{Y}_{N_2} \tag{2.88}$$

and depicted as logarithmic grayscale plot. For higher orders a part of the signal is mapped into the lower order components as aliasing components. It can be seen that these aliasing artifacts reach lower orders only for input values of higher orders. Similar to aliasing products in the time-frequency representation, a mirroring of the signal in the spectral domain occurs for this kind of sampling scheme.

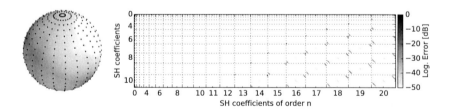

Figure 2.3.: Equiangular quadrature: Sampling scheme and aliasing map for precise quadrature with the maximum oder $n_{\max} = 10$.

2.5.2. Gaussian quadrature sampling scheme

The Gaussian quadrature sampling is (like the equiangular quadrature sampling) a rectangular sampling and thus well-suited for applications that use discrete steps in both angular directions. Its advantage to the equiangular quadrature sampling is the higher sampling efficiency of 50%, so only half the number of samples have to be used for the representations of functions with the same maximum order [Zot09a].

The continuous integral of the integration of continuous function is approximated the same way as with the equiangular quadrature (cf. Eq. (2.86)), using the constants of

$$L = n_{\max} + 1 \qquad \text{and} \qquad K = 2n_{\max} + 1. \qquad (2.89)$$

In Fig. 2.4 the sampling error of higher order spherical functions (plotted up to order 20) on a Gaussian quadrature sampling of maximum order 10 is depicted. The orthonormality matrix as defined in Eq. (2.88) is plotted after subtracting the main diagonal entries of perfect reconstruction. Compared to the equiangular quadrature only half the sampling points are needed for the same maximum orders, yielding slightly higher aliasing components. Functions given with a degree not higher than N_1 are represented without loss of information, while higher order components are mirrored to lower orders. The aliasing structure is similar to the equiangular quadrature and does not affect all orders. An input signal with an effective maximum order of e.g. $n_{\max} = 15$ only disturbs the resulting orders of $n \geq 7$ when using a Gaussian quadrature sampling.

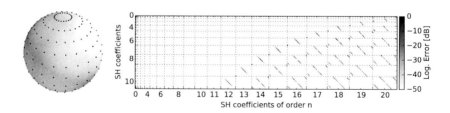

Figure 2.4.: Gaussian quadrature: Sampling scheme and aliasing map for precise quadrature with the maximum oder $n_{\max} = 10$.

2.5.3. Sampling scheme using hyperinterpolation

The sampling scheme using hyperinterpolation on the sphere is the most efficient representation of order limited functions in the spatial domain, having the best possible sampling efficiency of 100% [Zot09b]. For each coefficient of the SWS exactly one spatial sampling point is used for perfect reconstruction of order limited spherical functions. Details on the sampling distribution on the 2-sphere can be found in Sloan et al. [Slo99; Slo04].

Having an equal number of SH coefficients and sampling points the matrix $\mathbf{Y}$ becomes a square matrix. With its full rank, the matrix inversion can be performed using conventional matrix inversion, regarding a unique and precise operation without loss of accuracy for band limited functions [Zot09b; Zot09a]. $\mathbf{Y}$ thus describes the bijective transformation matrix converting SH coefficients to spatial data by multiplication. The inverse operation regards the SHT and can be expressed as

$$\hat{\mathbf{f}} = \mathbf{Y}^{-1}\mathbf{f}. \tag{2.90}$$

The precomputed matrix inverse of $\mathbf{Y}$ can thus be used for solving the SHT using simple matrix multiplication. The disadvantage of this kind of sampling scheme is the stronger spatial aliasing components that occurs for SH orders above order N_1. As there is no redundancy, these higher orders have a strong impact on the reconstruction of lower orders, as can be seen in Fig. 2.5. Again, the orthonormality matrix

$$\mathbf{O} = \mathbf{Y}_{N_1}^{-1}\mathbf{Y}_{N_2} \tag{2.91}$$

with the subtracted diagonal matrix for perfect reconstruction within the order limit is plotted.

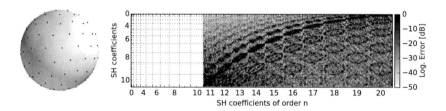

Figure 2.5.: Hyperinterpolation: Sampling scheme and aliasing map for precise quadrature with the maximum oder $n_{\max} = 10$.

2.5.4. Equiangular sampling schemes with regular spacing

Regular equiangular sampling schemes are commonly used for spherical measurements due to their algorithmic simplicity and the wide support in acoustic measurement software and spherical data formats[10]. Hereby a fixed angular resolution for azimuth and elevation is set and the measurements are performed in these steps. The resolution of the chosen sampling scheme can be stated in the $(\Delta\theta/\Delta\phi)$ format, with commonly used formats of $(10°/10°)$, $(5°/5°)$ or $(1°/1°)$. When starting and ending the elevation at the poles, the number of unique points of the chosen grid can be given as

$$I = \left(\frac{180°}{\Delta\theta} - 1\right)\frac{360°}{\Delta\phi} + 2. \tag{2.92}$$

The removal of duplicate points at the poles is essential for the calculation of the SWS, as the transformation matrix $\mathbf{Y}$ needs to have full rank. In practice,

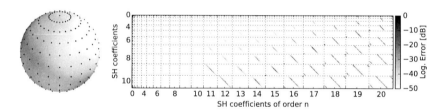

Figure 2.6.: Sampling scheme and aliasing map for $\Delta\theta = 15°$ and $\Delta\phi = 15°$ with the maximum oder $n_{\mathrm{max}} = 10$ using pseudoinverse.

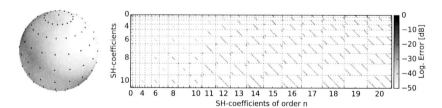

Figure 2.7.: Sampling scheme and aliasing map for $\Delta\theta = 20°$ and $\Delta\phi = 20°$ with the maximum oder $n_{\mathrm{max}} = 10$ using pseudoinverse.

---

[10]The measurement software Monkey Forest (MF), as well as the data formats OpenDAFF [Wef10] and the Common Loudspeaker Format (http://www.clfgroup.org) currently only supports this kind of sampling scheme.

duplicate measurement points can be used to quantify the uncertainty of the measurement system, as they should ideally yield perfectly identical results.

The matrix of the sampled spherical harmonic basis functions as defined in Eq. (2.69) can be inverted using a generalized inverse. Depending on the chosen maximum order, this yields to specific errors for functions with components below and above this limit. Exemplarily the errors for a chosen maximum order of 10 are depicted in Fig. 2.6 and Fig. 2.7, using a $(15°/15°)$ and a $(20°/20°)$ sampling scheme, respectively. It can be seen that while the former yields error free transformations for spherical functions consisting of SH orders of 10 or lower, the coarser sampling scheme shows errors also in lower orders. The choice of the maximal order is thus essential for processing an arbitrary sampling scheme, as the results will vary depending on that choice.

The orthonormality matrix for the calculation is computed using the generalized inverse as

$$\mathbf{O} = \mathbf{Y}_{N_1}^{+} \mathbf{Y}_{N_2} \qquad (2.93)$$

Contrary to the quadrature samplings all aliasing errors immediately affect low orders. While quadrature samplings gradually affect lower and lower orders for rising input orders, the regular equiangular sampling does not show an error free reconstruction for low orders in case aliasing occurs.

2.6. Measurement of acoustic transfer paths

Acoustical measurements allow to determine the impulse response of a given transfer path. The impulse response $h(t)$ of a linear time-invariant (LTI) system can be formulated in the time domain as a deconvolution of the recorded signal with the excitation signal, or as a spectral division in the Fourier domain.

For details on acoustical measurements refer to Müller et al. [Mül01]. Dietrich [Die13] describes the total electro-acoustical signal path with a complete measurement chain that can be extended with the directivity patterns of source and receiver.[11]

The general characterization of acoustic systems require the components to be linear and time-invariant (LTI). In reality often weakly nonlinear system or com-

---

[11] All measurements have been performed using the *ITA-Toolbox* for Matlab, cf. also Appendix A.4.1.

ponents occur, as well as time-variances (especially with room acoustical measurements) that create a specific uncertainty in the measured results [Die13].

2.6.1. Multiple exponential sweep method

In the context of head-related transfer functions Majdak et al. [Maj07] describe a technique called *Multiple Exponential Sweep Method* (MESM). In the measurement multiple loudspeakers are used simultaneously for the measurement. The signals of the output channels are timed so that the excited frequencies and their nonlinear harmonics do not overlap in order to retrieve the full information. The multichannel data measured in this work uses an optimized version of the MESM as described by Dietrich et al. [Die12b]. Hereby, weak nonlinearities are accounted for and do not influence the obtained results at the same or at another channel.

Measurements using the MESM are especially beneficial for experiments involving humans beings. As the measurements can be performed faster, the discomfort for the subjects can be minimized and the deviations from movements minimized.

2.6.2. Reciprocity of transfer paths

The principle of reciprocity is commonly found in textbooks about wave propagation or linear-circuit theory. According to Beranek [Ber54] the principle of reciprocity can be extended to transducers, that are coupled by a gaseous medium. He states that such a transducer fed by a constant-current generator produces a given open circuit voltage at a second transducer. The same current applied to the second transducer induces the identical open circuit voltage at the first transducer. The ratio of resulting voltage and excitation current is thus constant as

$$\frac{u_2}{i_1} = \frac{u_1}{i_2}. \tag{2.94}$$

Consequently, the measurement of acoustic transfer paths can be performed in both directions and can alternatively be solved by the identical setup with sender and receiver interchanged. Reciprocity is used in order to simplify boundary value problems as formulated in Sec. 2.3.4. One example of such a simplification is the incoming wave towards a human listener. Applying the principle of reciprocity, the initial problem of having sources in the far-field and additional scattering sources in the vicinity of the human ear, is simplified to a exterior problem: All sources (one at the location of the microphone and many scattering sources) can be confined to a limited area with no additional sources outside that area, yielding a classical exterior problem (cf. also Sec. 4.3). While at arbitrary field points the interchange of source and receiver generally creates a different acoustic field, the signal picked up at the receiver is identical and thus yields the identical transfer path [Dur04].

3

Directivity patterns of natural sound sources

In this chapter the measurement, processing and analysis of directivity patterns of natural sound sources are described. Due to the lack of precisely repeatable excitation, directivity patterns of natural sound sources are usually measured simultaneously for all directions using large surrounding spherical microphone arrays. Directivity patterns are commonly implemented as logarithmic measure derived from averaging a set of recorded tones [Mey78; Mar85; Oto04; Len07]. Phase information is hereby not taken into account.

In order to obtain complex-valued spectra representing the directivity patterns of musical instruments, a large spherical surrounding microphone array is employed that encompasses the musician who is placed in the geometric center of the array. It allows the simultaneous recording of the emitted sound in all directions, which is used to retrieve phase accurate directivity patterns of these natural sound sources. Tonal sound sources radiate a set of partial tones, consisting of the fundamental frequency and its higher harmonics, which are excited simultaneously for each played pitch. The recordings are analyzed in order to derive information about similar and dissimilar directivity patterns that are encountered for specific frequency intervals.

The knowledge of the complex directivity patterns furthermore allow an analysis in terms of the physical origin of the emitted acoustic waves. As the source position has significant impact on the results of a spatial Fourier analysis, alignment algorithms can be applied in post-processing in order to match the center of the sound radiation to the geometric center of the surrounding spherical array. These acoustic centering approaches allow to derive more compact representations of the directivity patterns in terms of their spherical wave spectra and have the potential to suppress aliasing at certain frequencies.

Parts of this chapter have been published in [Pol12e] and [Beh12].

3.1. Recording of musical instrument radiation patterns

For the discussion of directivity patterns of natural sound sources a clear terminology is beneficial: A single *tone* played by a musical instrument has a perceived *pitch* of the *fundamental frequency* of the tone. Each tone consists of a set of *partial tones*, the fundamental and the set of *higher harmonics* occurring at frequencies of integer multiples of the fundamental frequency. All tones that can be excited of a specific instrument are regarded as its *gamut*.

As mentioned in Sec. 2.1 several definitions for the directivity pattern exists, each suiting specific demand. For the application of auralizing a specific sound recording performed at a known direction from the instrument, the *directivity factor* with that particular reference direction can be used in order not to colorize the recording in an artificial manner.

For practical applications it is useful to be able to use arbitrary dry recordings[1] of musical instruments in order to combine them with the directivity pattern of these instruments. The source material can then be changed flexibly while using a set of directivity patterns that have been derived earlier. This approach is valid if the following assumption can be made: The directivity patterns are considered as a frequency dependent pattern without any significant change due to the style and the strength of playing, as this information is usually not directly included in recording used for auralization.

In reality identical or similar frequencies can be excited from tones of various pitches with their corresponding set of partial tones. The pitch of a specific tone, e.g., has the same fundamental frequency as the first harmonic partial tone of a pitch one octave below. The directivity patterns at specific frequencies excited from differing pitches do not necessarily show high similarity. In order to quantify the deviation from an averaged directivity pattern, the complex patterns of the different harmonics for all possible pitches are compared for various instruments.

3.1.1. Types of sound excitation

The term *musical instrument directivity pattern* can be defined in different ways: Natural sound sources can be excited artificially by the use of a technical apparatus that mimes the excitation by a musician. Artificial excitation is generally

---

[1] A perfectly dry recording captures only the direct sound of the acoustic source without reflections from the recording environment. These recordings are commonly performed in an anechoic chamber for best possible results.

repeatable and thus allows to obtain high resolution data using sequential measurements helping to understand the physical properties of the musical instrument, as done by Cremer [Cre81] and [Bad05].

In a real scenario, however, the musician is always present and regards an integral part of the complete arrangement. The directivity pattern of the player with the musical instrument is signifiant for possible applications in room acoustics. In order to obtain these patterns, either the musician is supposed to be present during the measurement, or (when measured without player using artificial excitation) should be added to the result in post-processing. This deviation could possibly be approximated by a modification of the measured directivity pattern, implementing an object with the rough shape of the musician in the vicinity of the musical instrument.

More realistic results can be expected measuring the radiation of musical instruments using natural excitation by the musician. The sound diffraction around the human body and the scattering is fully included in the measurements using this setup. Furthermore, natural excitation is the most natural form of sound production as encountered in musical performance. As exact reproducibility cannot be accomplished for natural excitations, surrounding microphone arrays are commonly employed to capture the sound pressure values simultaneously in all directions.

3.1.2. State of the art

In literature different strategies for the measurement of sound radiation can be found, using natural excitation with simultaneous recording of all directions and using artificial (repeatable) excitation in combination with a sequential measurement approach. One of the earliest works regarding this research question is the series of articles published by Meyer about half a century ago [Mey64; Mey65a; Mey65b; Mey65c; Mey66a; Mey66b; Mey66c; Mey67]. These directivity pattern have been obtained with the help of a turntable and exciter and can be found in comprised form in [Mey78]. An alternative method to achieve repeatability is the use of reciprocity, by swapping sender and receiver position when measuring transfer paths (cf. Sec. 2.6.2). Weinreich [Wei97] applies the principle of reciprocity by using a technical source at a distance from the instrument and measuring the force as it occurs on the bridge of a violin.

More recently Otondo et al. [Oto04] published results of measured directivities, recorded simultaneously with 13 microphones distributed on the median (vertical) plane and the horizontal plane, and analyzed the data using magnitude averaging. In the same year Slenczka [Sle04] measured the directivity patterns using a microphone array of 24 loudspeakers distributed spherically around the musician (see Appendix A.1.3 for the geometry), using a similar approach by using a geometric average of the spectra by recording a complete piece of music. This data has been recorded with the purpose to do auralization in the CAVE-like environment of RWTH Aachen University [Len07].

In 2009 Hohl [Hoh09] measured directivity pattern using 64 microphones solidly arranged using the hyperinterpolation sampling scheme (cf. Sec. 2.5.3) for a maximum SH order of 7 in order to perform complex (phase accurate) analysis of the measured data. In the same year the directivity patterns and sound power spectra of a large set of symphonic musical instruments were measured in a joint-project of TU Berlin and RWTH Aachen University. Preliminary results of the data processing were published in [Pol09b] and [Pol10b]. Pätynen et al. [Pät10] published an analysis of several directivity patterns of symphony orchestra instruments using cross-sections of the averaged directivity pattern obtained with 22 microphones, all of them using natural excitation of the instruments.

3.1.3. Measurement equipment

In recent years a large surrounding spherical microphone array has been developed at the Institute of Technical Acoustics of RWTH Aachen University. It consists of a lightweight fiberglass structure arranging 32 microphones spherically in the direction of the faces and vertexes of a dodecahedron, located at an (almost) constant radius from the geometric center. The geometry of the array is depicted in Fig. 3.1 with the locations of the microphones openings (*Sennheiser KE4-211-2*) with respect to the geometric center of the microphone array listed in Appendix A.1.4.

In Fig. 3.2 the placement of a musician inside the spherical array is depicted in the hemi-anechoic chamber of RWTH Aachen University. In order to provide for a flexible placement of the musician the used chair can be placed arbitrarily in the horizontal plane and has an adjustable height. When possible, the assumed acoustical center of the musical instrument was aligned roughly to the geometric center of the spherical array (so a higher position than shown in Fig. 3.2 was usually chosen for the recordings).

All microphones are fully calibrated as described in Appendix A.2.1. Additionally the directivity patterns of the array sensors are relevant if measuring objects with

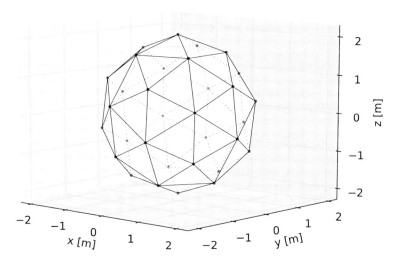

Figure 3.1.: Geometry of the spherical microphone array used for musical instrument directivity recording

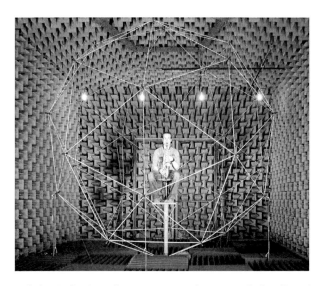

Figure 3.2.: Spherical microphone array used to record the directional dependent radiation of natural sound sources

a significant spatial expansion. It is desired that the effective sensitivity of each sensor is not influenced by the angle of sound incidence. In Appendix A.2.2 two setups for the microphones as used in the surrounding spherical array are shown and analyzed, concluding that the deviations in the meaningful part of sensor directivity can potentially distort the results significantly.

3.1.4. Measurement environment

The recording of directivity patterns of musical instruments obey the same requirements as conventional acoustic measurements where any reflections from obstacles have to be avoided. Contrary to measurements using deterministic measurement signals, an elimination of late reflections is not possible when recording natural sound sources.

Using the surrounding microphone array in the hemi-anechoic chamber shows interference effects in the resulting directivity patterns, depending on frequency and elevation angle. This is caused by the ground reflection that is usually undesired in the results of the measurements. Practical experiments in the hemi-anechoic chamber showed, that the use of acoustic absorbers covering the solid floor of the chamber yields insufficient suppression of the reflection occurring on the ground. One layer of absorbers of an approximate height of 20 cm exhibits a reflection factor of 0.5 or higher for frequencies below 400 Hz, making the setup inadequate for the desired accuracy and frequency range [Beh08]. Stacking up several absorbers to a larger thickness improves the quality while diminishing the available size for the measurement array. As the edges of the absorbers also cause undesired secondary sources, the use of a full anechoic chamber is advisable for high quality measurements.[2]

3.1.5. Obtaining recordings for a directivity database

In collaboration with the Audio Communication Group of TU Berlin the measurement data used for the directivity database have been obtained in the full-anechoic chamber located at the Institute for Technical Acoustics of TU Berlin. This room has a low cut-off frequency of 63 Hz, delivering sufficient attenua-

---

[2] As the constructed microphone array only fits the hemi-anechoic chamber of the Institute of Technical Acoustics of RWTH Aachen University (the full anechoic chamber is not sufficiently large), all measurements used in this thesis have been performed in external measurement facilities in collaboration projects.

tion in the frequency ranges of interest and facilitates a high quality analysis of measured musical instrument directivity patterns.

The recording of clean tones for the directivity database has been performed by placing the musicians individually in the center of the surrounding spherical microphone array. Both modern and historic (mostly symphonic) instruments were employed, a detailed list of the instruments can be found in Appendix A.3.1. A set of three tones for each pitch played (usually chromatic scales) in *pianissimo (pp)* and *fortissimo (ff)* has been recorded with a manual selection of the cleanest tone chosen for the subsequent analysis.[3]

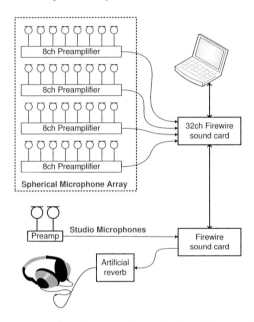

Figure 3.3.: Setup used for the recording of directivity patterns of natural sound sources

In Fig. 3.3 the schematic setup as used for the recording is depicted. The 32 microphones are connected via four 8-channel preamplifiers in the anechoic chamber and connected to a Firewire sound card. A second sound card is connected to the recording PC in order to provide enough channels for an additional high quality recording of the sound. Artificial reverb allows the musician to listen to

---

[3]The measurement equipment was fully calibrated as described in Appendix A.2.1 without the optimizations as described in Appendix A.2.2.

their own sound with sufficient reverberation to feel comfortable in this rather unnatural environment.

3.1.6. Recording audio for auralization

Beside the recording of clean tones it is also possible to record specific audio tracks for a multichannel sound recording including the directivity pattern of the instruments. The knowledge of a static directivity database is not sufficient for this application if dynamic movements are meant to be included. For these multichannel recordings for the purpose of auralization the same measurement setup can be used together with additional high quality microphones to increase the quality of the recordings in terms of signal-to-noise ratio (SNR). The microphones of the spherical array have a higher self-noise and are thus used only for the determination of the directivity, not as source material for auralization. For the application of measured directivity patterns onto a recording performed at a single direction (r, θ_0, ϕ_0) the definition of the directivity factor as defined in Eq. (2.1) is useful.

An example for such a recording session is the time synchronous recording of a medieval choir in cooperation with the Polytechnic University of Madrid, Spain, using 32 microphones for the directivity measurement and two studio microphones for the auralization.[4] In Fig. 3.4 the complete setup as used in the recording in Madrid is depicted. Due to the limited height of the anechoic chamber the southernmost microphone approximately half a meter below the feet of the singer has been omitted, resulting in a total of 31 channels for the directivity measurement. Using the headphones the singers have been provided with a mix of their live recording with added reverb and the audio tracks of the videos of a full choir performing have been played on the iPad simultaneously.

3.2. Processing of recorded data

The multichannel tracks of the sound recordings can be used for further processing. Here several types of computations are possible, depending on the desired ap-

---

[4]"The musical pieces recorded have been performed by members of the musical group Schola Antiqua. [...] A total of eight pieces of the Mozarabic Chant repertoire have been recorded. Seven of them belong to the Office of the Dead, and the eighth to the Rite of Consecration of the Altar. In order to characterize different vocal timbres, the recording has been made by six different singers. Each of them has played all the musical pieces chosen." [Ped12] The goal of the project is the auralization of the musicians in a simulated room acoustical environment of an ancient church, including dynamic movements of the singers.

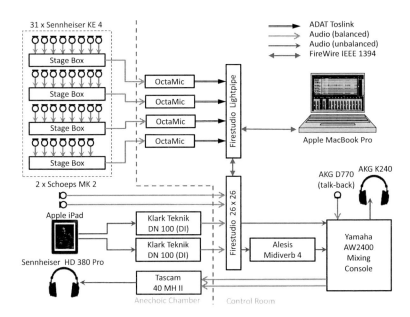

Figure 3.4.: Setup used for the synchronous recording of audio with directivity patterns [Ped12]

plication and the required data format. Zotter [Zot09a] lists many techniques for directivity processing of source directivity measured with a surrounding spherical microphone array and distinguishes between time-frequency, space-frequency, space-time-frequency domain methods, with some of them reviewed in this section.

3.2.1. Analysis of the used array geometry

As shown in Sec. 2.5 the distribution of the microphones can be analyzed in terms of aliasing components created by higher orders. In Fig. 3.5a and Fig. 3.5b the aliasing error for directivity patterns up to a SH order of 10 for sources located in the center of the used spherical microphone array is visualized. The DSHT is calculated for these plots using the pseudoinverse matrix of $\mathbf{Y}$ that is defined in Sec. 2.4.1.

As the spherical microphone array consists of 32 sensors, the maximum order n_{max} for the computation is set to have roughly an equal number of sensors and SH basis functions, using $n_{max} = 4$ (for a total number of 25 spherical harmonics) and $n_{max} = 5$ (for a total number of 36 spherical harmonics). As can be seen in Fig. 3.5a a function with components in order 5 yields comparatively small distortions in lower (odd) SH orders. Calculating with a maximum order of 5 instead (Fig. 3.5b) yields higher distortions that are limited to the same order for a directivity pattern with a component of order 5. The grayscale in the figure represents the logarithmic error component of a specific SH coefficient with respect to the excitation level.

Note that although the aliasing structure in this plot looks similar to that of a quadrature sampling (cf. Fig. 2.4), a directivity pattern of order 6 yields components both in order 4 and in order 0 (monopole part) with the given sampling scheme of the employed spherical microphone array (independent of the assumed maximum SH order). These aliasing errors are unavoidable for higher orders of directivity patterns encountered.

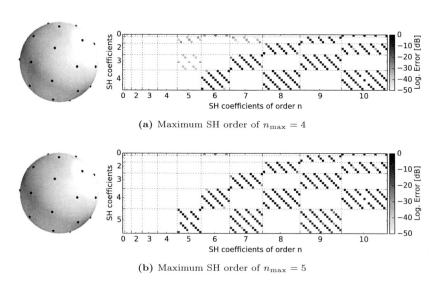

(a) Maximum SH order of $n_{max} = 4$

(b) Maximum SH order of $n_{max} = 5$

Figure 3.5.: Sampling scheme and aliasing map for surrounding spherical microphone array implementing the DSHT by pseudoinverse with different maximum order.

3.2.2. Averaged magnitude data

Current room acoustical simulation software use mostly sound pressure levels at discrete directions data as input parameters [Oto04; Sch11]. As the resolution of musical instrument measurements is usually rather low, a spatial interpolation on the magnitudes values in frequency domain is commonly performed [Len07].[5]

The source material for deriving an averaged directivity can either be chromatic scales or any arbitrary piece of music with a representative pitch coverage. The directivity is obtained by taking the Fourier transform of the time signal of all recorded channels (either on the whole piece or segmented into smaller time-blocks) and performing the desired band averaging (e.g. in third band intervals) on the magnitude spectra derived by applying the discrete Fourier transform (DFT) on the tracks recorded by the microphones. The spatially discrete data can then be interpolated to a higher resolution, e.g. by representing the directivity as spherical wave spectrum as described in Sec. 2.3.3. For the auralization of sound sources with directivity, the recording channel is normalized by the directivity pattern at the direction of the reference microphone in the dry recording not to colorize the spectrum. Applying the directional and frequency dependent filter, a dry audio passage can be used to account for the levels differences of the sound radiation.

The processing using averaged magnitude data implicitly assumes that directivity patterns of natural sources are uniquely defined for a specific frequency with a smooth transition between adjacent frequencies. In Sec. 3.3 it is shown for which instruments this prerequisite is fulfilled.

3.2.3. Time-dependent analysis

Contrary to time averaging the data can also be processed as a time-dependent quantity for a particular piece of music, allowing to include natural movements of the musicians. These dynamic directivity patterns find their applications in the field of virtual reality. The motion capturing of musicians as performed by Schröder [Sch11] can employ dynamic directivity patterns[6] in order to perform

---

[5]RAVEN, a room acoustical simulation software developed by Schröder [Sch11] the Institute of Technical Acoustics at RWTH Aachen University, uses recorded data stored in averaged third band octave bands and interpolated on a regular equiangular grid of $10°$ resolution in both azimuth and elevation (cf. Sec. 2.5.4). For fast retrieval of the directivity data at a specific direction the data is stored in the OpenDAFF file format [Wef10].

[6]Alternatively these dynamic patterns can be achieved artificially by using a static directivity database and performing rotation either in spatial domain or in the SH domain by using the Wigner-D rotation matrix, cf. Sec. 2.3.5.

auralization of dynamically changing sources. Also changes of the dynamics or the style of playing can be taken into account when using time-variant directivities.

The auralization of audio content recorded in the surrounding spherical microphone array can be done using blockwise processing. The time-variant processing of directivity patterns slices the recorded multichannel audio tracks in time frames of arbitrary length and calculates the frequency dependent directivity patterns for each block. As the spectrum of a played tone regards a line spectrum and is thus sparse in the frequency domain, the directivity pattern is not defined for the continuous spectrum. In combination with the simultaneously recorded audio track only the excited frequencies are required in that time frame. Using this method allows the use of affordable microphones for the array and (at least) one single high-quality microphone recording the track used for the auralization. This type of data processing has been applied in collaboration with Pedrero et al. [Ped12].

3.2.4. Harmonic peak extraction

A third type of processing is the determination of complex directivity patterns (consisting of the complex spectral values of all occurring partial tones) in order to use them for a profound analysis of the radiation of natural sound sources. The source material used are the discreetly recorded tones, covering the full gamut.[7]

Each recorded tone is processed individually by extracting the time interval of steady excitation and performing a Fourier transform on the signals. In order to avoid spectral leakage a window is applied over the complete steady part before performing the FFT.[8] The resulting line spectra are sampled at the peaks of the partial tones and stored as amplitudes with phase relations for all microphones of the spherical array. As the amplitude peaks of the partial tones show a rapidly changing phase around their physical resonance, the phase has to be determined considering a small frequency interval around the peak frequency. Instead of using the comparatively low spectral resolution of the data (due to comparatively short patches of steady tone excitation) the phase transition can be interpolated from the spectrally sampled phase information. As the knowledge of absolute phases are not required, a common data processing for all microphone channel

---

[7]The gamut of modern instruments usually encompasses several octaves in a chromatic scale.

[8]The type of window used is not significant as the effect of the window on the spectrum regards to all channels. In the data used for this work a Hann window was applied.

allows to derive correct phase relations. A flowchart of the processing is depicted in Fig. 3.6.

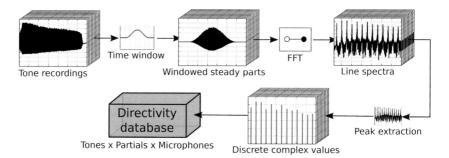

Figure 3.6.: Process chain for the recorded tones to obtain a directivity database with complex peak information of the all partial tones (fundamental and higher harmonics) a musical instrument can excite.

The obtained dataset contains complex values that quantify the magnitudes and phase relations of the higher harmonics with respect to the fundamental frequencies. For each partial tone of all pitches that a musical instrument can excite a spatially discretized directivity pattern as defined in Eq. (2.1) is derived (with a reference direction of high level that can be chosen from the set of discretely measured direction). These data can be used for deriving the acoustic center of the spherical wave emitted (cf. Sec. 3.4). To add dynamics to the static directivity patterns the movements of the musicians can be obtained and recorded by motion tracking methods [Sch11].

The number of partial tones used for spectral averaging (as described in Sec. 3.2.2) varies with the pitch and the used type of filter. In Fig. 3.7 the frequencies of the first ten partial tones for all pitches of a single musical instrument (English Horn) are depicted. The fundamental frequencies of that instrument range from approx. 150 Hz to almost 1 kHz, with the higher harmonics at multiples of the fundamental frequency. Exemplarily a third octave band interval at 500 Hz and an octave band interval at 2 kHz are marked in red and blue, respectively, showing the number of partial tones that are taken into account in an averaging process. It can be seen that the interval at 500 Hz contains partials from the lowest three pitches, while the octave band interval at 2 kHz contains a much higher number of partial tones of mostly higher order.

Depending on the similarity of the directivity patterns within a specific band interval, the directivity patterns of the specific partial tones can be represented more or less accurately by an averaged frequency dependent directivity pattern as described in Sec. 3.2.2. As the phase relations between different played pitches are unknown, averaging approaches using the complex spectra have no unique result for dissimilar directivity pattern.[9]

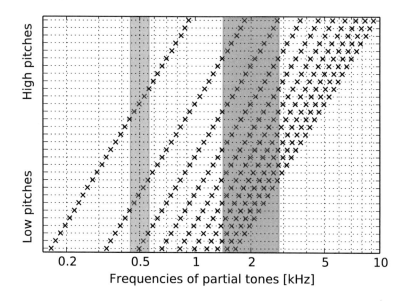

Figure 3.7.: Frequency map of the first ten partial tones for all recorded pitches of a instrument (here plotted for the English horn). Third octave band interval (red) at 500 Hz and octave band interval (blue) at 2 kHz marked exemplarily.

3.3. Analysis of spectral smoothness

As the wave number (and thus the wave propagation term) changes over frequency, directivity patterns naturally vary with frequency. The directivity thus is often considered as frequency dependent function independent of played pitch,

---

[9]Using one microphone as reference direction for the pressure spectra, the arbitrary phase relations between different played pitches can set to zero at the reference channel. The choice of reference, however, influences the result of the averaging approach, so working with magnitude values might be the better choice when determining an averaged directivity.

volume or playing style. Depending on the type of sound source this is an approximation to a greater or lesser extent. The processing as described in Sec. 3.2.2 neglects these differences and extracts a *generic directivity pattern* with one defined spherical function for a given frequency. As can be seen in Fig. 3.7, each center frequency of an interval used for averaging possesses a number of partial tones of different pitches, possibly each with a specific directivity pattern, even though these different partial tones radiate in the same frequency ranges.

The extracted harmonic peak values allow to quantify the deviations of the directivity patterns within a specific band interval. Dissimilar directivity patterns of partials within the same band, yield higher deviations to the generic pattern derived by averaging.

In order to compare the directivity patterns of all pitches with their corresponding partial tones, the normalized correlation coefficient as defined in Eq. (2.59) is calculated as a measure of similarity for all possible pairwise combinations of partial tones an instrument can emit. The definition of the correlation coefficient and the squared magnitudes as given in Eq. (2.58) and Eq. (2.60), respectively, adjusted for the array geometry using $L_{\mathrm{mics}} = 32$ sensors can be given as

$$C(\mathbf{p}_1, \mathbf{p}_2) = \frac{4\pi}{L_{\mathrm{mics}}} \mathbf{p}_1{}^H \mathbf{p}_2 = \frac{4\pi}{L_{\mathrm{mics}}} \sum_{i=1}^{L_{\mathrm{mics}}} \overline{p_{1,i}} \cdot p_{2,i}, \tag{3.1}$$

$$E_1 = \frac{4\pi}{L_{\mathrm{mics}}} ||\mathbf{p}_1||_2^2 = \frac{4\pi}{L_{\mathrm{mics}}} \sum_{i=1}^{L_{\mathrm{mics}}} |p_{1,i}|^2 \qquad \text{and} \tag{3.2}$$

$$E_2 = \frac{4\pi}{L_{\mathrm{mics}}} ||\mathbf{p}_2||_2^2 = \frac{4\pi}{L_{\mathrm{mics}}} \sum_{i=1}^{L_{\mathrm{mics}}} |p_{2,i}|^2 \tag{3.3}$$

with $\mathbf{p}_x = (p_{x,1}, p_{x,2}, \ldots, p_{x,L_{\mathrm{mics}}})^T$ being the complex pressure values of the partial tone x at all microphones and $||\mathbf{p}_x||_2$ being the 2-norm of this vector. The normalization factor $\frac{4\pi}{L_{\mathrm{mics}}}$ is applied in order to scale the sum at the microphones to the result of an continuous integration over S^2 (surface of unit sphere).

Written in vector form the normalized correlation coefficient can thus be expressed as

$$\widetilde{C}(\mathbf{p}_1, \mathbf{p}_2) = \frac{C(\mathbf{p}_1, \mathbf{p}_2)}{\sqrt{E_1 E_2}} = \frac{\mathbf{p}_1{}^H \mathbf{p}_2}{||\mathbf{p}_1||_2 \cdot ||\mathbf{p}_2||_2} \tag{3.4}$$

yielding a complex value expressing the similarity of the measured directivity patterns. As the phase relation between the pairs of these patterns is arbitrary, the magnitude value of the normalized correlation is used to quantify the similarity of complex directivity patterns on a scale from zero to one.[10]

3.3.1. Plotting cross-correlation values over frequencies

The normalized correlation values of all pairs of partial tones (using all recorded tones of a single musical instrument) are analyzed in a diagram with the frequencies of the two partial tones used as logarithmic location on the x- and y-axis.[11] This yields a scatter plot showing the correlation as color information of all pairs of partial tones, the pair with identical frequencies plotted at the main diagonal.[12] Using logarithmic scales for both frequencies, all frequency pairs with fixed intervals are mapped onto lines. The structure of the plots is shown in Fig. 3.8 with the lines for third octave band intervals and octave band intervals marked with a dashed or dotted line, respectively.

The correlation values are plotted in the lower right triangle in rising order with respect to their magnitudes as opaque dots (thus showing the highest correlation values for any pair of partial tones with the given frequencies) and in the upper left triangle in falling order (showing the lowest correlation values). Comparing these two triangles gives a hint about the spread of the correlation values at identical frequency pairs of the analyzed partial tones. A strong deviation between the correlation values in the two triangles show that some partial tone combination of that specific frequency pair have rather similar directivity patterns, while other tone combination (of identical frequencies) radiate sound in a fundamentally different manner.

The areas marked in red and blue regard the areas of consideration for third octave band and octave band averaging. The range of fundamental frequencies of all analyzed tones are marked with gray tags at the upper edge of the plot (in the legend the fundamentals of the English Horn are depicted, cf. Fig. 3.7).

For N_t tones, and N_p partials per played tone analyzed $(N_t \cdot N_p)^2$ normalized cross-correlation values are calculated. For an average instrument with a gamut of approx. three octaves more than 100.000 correlation values have to be computed

---

[10]Note that although the (single-value) phase relation between a pair of directivity patterns is not taken into account, a non-constant phase difference on the sphere between the different partial tones is detected and punished with a lower correlation value.

[11]The size of the markers is adjusted to the density of points in that area in order to obtain only little overlap of the point, but large enough point sizes for convenient visual comparison of the data sets.

[12]The correlation values for identical tones are unity, provide no information and have been removed from the plots.

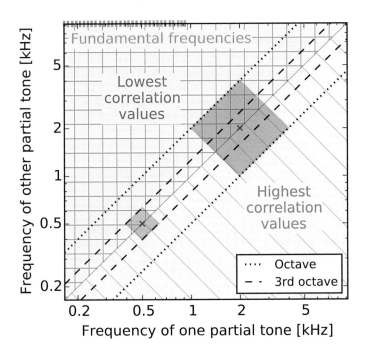

Figure 3.8.: Legend for the correlation plots of all partial tone pairs of a musical instrument. In the upper left triangle the lowest correlation values are plotted opaquely on top, in the lower right triangle the highest correlation values are dominant.

for the scattering plots of the correlation values.[13] The number of sampling points typically used in surrounding spherical microphone arrays usually does not allow to gain sufficient information for obtaining the SWS for higher frequency where a higher variation occurs. The correlation values computed with that limited (and for higher frequencies most likely undersampled) set of sensors still predict the similarity of that functions, as long as the musician did not change his position during the time of recording.

The analysis depicted in Fig. 3.9 to Fig. 3.11 show generally higher values for the correlation values for partial tones at low frequencies. No significant difference has been found between the recordings obtained in *pianissimo (pp)* and

---

[13]Comparable plots can be found in Behler et al. [Beh08] and Baumgartner et al. [Bau10] for the analysis of similarity of directivity patterns of musical instruments.

fortissimo (ff).[14] As the recordings with high amplitudes show generally less noise influences and seem thus cleaner, the fortissimo data has been used for the processing. The first 10 partial tones for every pitch have been used for the analysis.

3.3.2. Woodwind instruments

In Fig. 3.9 the correlation values for the woodwind instruments (English Horn, Alto Saxophone, Tenor Saxophone, Oboe and Clarinet) are depicted. For low frequencies below approx. 300 Hz to 400 Hz the correlation of all partials within this frequency range is high for all woodwind instruments. These instruments have in common a relative sudden change in the correlation values for rising frequencies. The mentioned effect can be observed well in the example of the *English Horn (cor anglais)*: All partial tones below 380 Hz show high correlation values, while the partials in the interval from 380 Hz to 520 Hz also show high correlations. The correlation of partials in-between these two groups, however, show remarkable low correlation values. This effect can be seen in all evaluated woodwind instruments (with the edges between the discrete intervals being more pronounced when analyzing less than ten partial tones for each pitch). The existence of correlation groups is expected to be caused by the abrupt change of the fingering for specific tones that are played with the technique of overblowing (a particular fingering can excite a higher pitch when overblowing the tone).

For frequencies above approx. 600 Hz the correlation values of all woodwind instruments show a higher spread, noticeable in the difference of the highest correlation values (in the bottom right triangle) and lowest correlation values (in the top left triangle of the plot). It can thus be concluded that the difference to a generic directivity derived by averaging becomes larger for rising frequencies. For the saxophones the correlation values of the directivity patterns are relatively high when considering averaging in third bands, while for the other woodwind instruments even the partial tones within the lines marking the third band interval do not show high similarity for higher frequencies.

3.3.3. Brass instruments

Brass instruments differ from woodwinds by having an immutable geometry of the parts responsible for the sound radiation for different pitches played. As can

---

[14]At fortissimo higher amplitudes in the higher harmonics have been detected, but for the calculation of directivity this spectral coloration is not relevant.

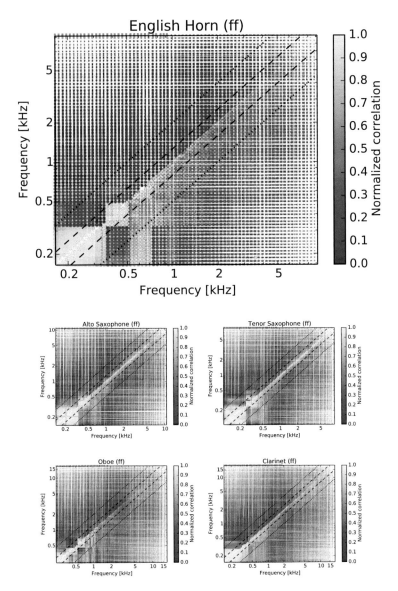

Figure 3.9.: Woodwind instruments: Normalized correlation of different partial tones of given frequencies. The dashed lines mark the 3rd octave band, the dotted lines the octave band interval. Fundamental frequencies marked with gray tags on top edge.

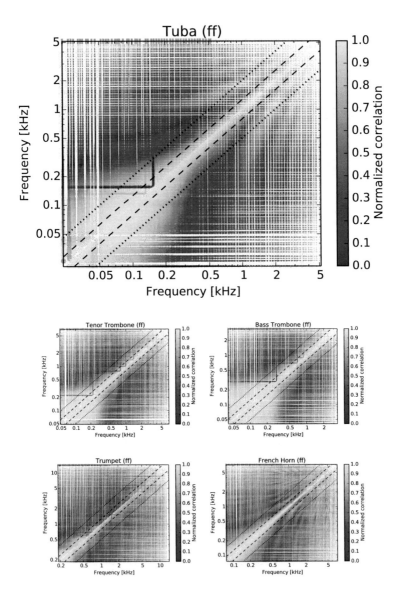

Figure 3.10.: **Brass instruments:** Normalized correlation of different partial
tones of given frequencies. The dashed lines mark the 3rd octave
band, the dotted lines the octave band interval. Fundamental
frequencies marked with gray tags on top edge.

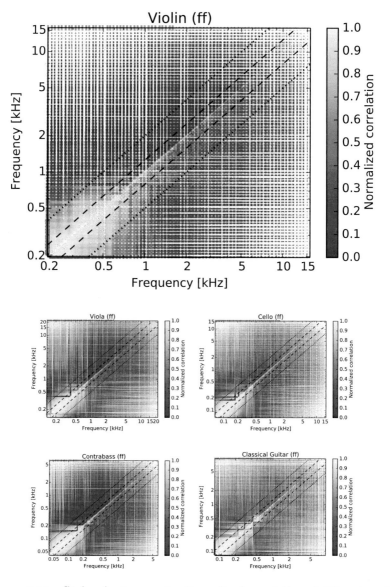

Figure 3.11.: String instruments: Normalized correlation of different partial tones of given frequencies. The dashed lines mark the 3rd octave band, the dotted lines the octave band interval. Fundamental frequencies marked with gray tags on top edge.

be seen in the plots in Fig. 3.10 these static geometries cause a rather smooth directivity pattern over frequency with low variations in both triangles of the plot (except for outliers). For frequencies above 300 Hz a third band averaging seems appropriate, as the correlation in that range stays at a high level up to high frequencies. The trombones even show relative high values within octave bands.

3.3.4. String Instruments

String instruments show generally the smallest correlation values between different partial tones as can be seen in Fig. 3.11. While for wavelength larger than the dimensions of the instruments the correlation is generally high, the correlation values drop steeply for higher frequencies. Even partial tones within a small frequency interval show low correlation values, suggesting a low similarity of the directivity. A band averaged generic directivity does not resemble well the individual directivity patterns encountered at a specific partial tone played by a sting instrument. The influence on the human perception regarding directivity patterns and the effect of using averaged directivity data instead of patterns for specific partials is still an open research question.

3.3.5. Statistical evaluation

Beside graphical evaluation of the correlation values of all partial tones for musical instruments, also statistical means can be employed. All pairs of partial tones with an interval smaller than a third octave interval or an octave interval are used as input values for a statistical analysis with the resulting plots shown in Fig. 3.12. These boxplots visualize the magnitudes of the normalized cross-correlation values located in a frequency range within the red (for third octaves or 4 semitones) or the blue square area (for octaves or 12 semitones) as depicted in the legend in Fig. 3.8.

These areas are being moved along the main diagonal in order to derive the frequency dependent correlation values. A minimum of six directivity patterns for each interval is required to display data in that frequency range. The median values are marked with a red or blue line, while the box extends from the lower to the upper quartile. The whisker limits are defined as the 1.5 times the interquartile range above the upper, or below the lower quartile [Tuk77].

The English horn as an example for the woodwind instruments shows very high correlation values for low frequencies with an abrupt decline for rising frequencies at around 300 Hz to 500 Hz. The discrete patches of high correlation as can be seen in Fig. 3.9 cause very spread boxes at frequencies around 300 Hz to 400 Hz, while for frequencies slightly above that interval the correlation values are generally high again. The octave band averaging does not show this rise to high values, as the correlation values generally decrease already for lower frequencies.

From the correlation plots in Fig. 3.10 the well-behaved nature of the directivity patterns of brass instruments can be seen. This also shows in the statistical analysis. The tuba as an example for the group of brass instruments shows high similarity of all directivity patterns (except for the outlier below 200 Hz) up to high frequencies. Third band averaging yields significantly higher correlation values than the octave band averaging.

The analysis of the violin as candidate for the string instruments shows high correlation values up to frequencies of approx. 600 Hz to 800 Hz and very spread values above that with a rather low median value. This confirms the observation that an averaged generic directivity pattern is not very similar to the actual directivity patterns of the partial tones of the instrument for wavelengths larger than the dimensions of the instrument.

3.4. Acoustic centering methods

The complex analysis of the recordings made with the surrounding spherical microphone array is strongly dependent on the chosen focal point for the multipole expansion. This point is located at the origin of the spherical coordinate system that is used for the SHT and commonly chosen to be located in the geometric center of the microphone array. In reality the sound emitted by a musical instrument is not bound to a single point and the contributing sources are not necessarily located in the geometric center of the microphone array.

As the translation of a centered sound source causes higher harmonics (cf. Sec. 2.3.5), the placement of the musicians with their instruments is a significant factor for the complex analysis of the directivity patterns and the compact representation in terms of their SWS. The representation for a non-optimal focal point regards an alternative representation and is not per se wrong. It requires, however, a considerable higher number of SH orders for accurate representation, so aliasing errors are more likely for a displaced source in a given sampling

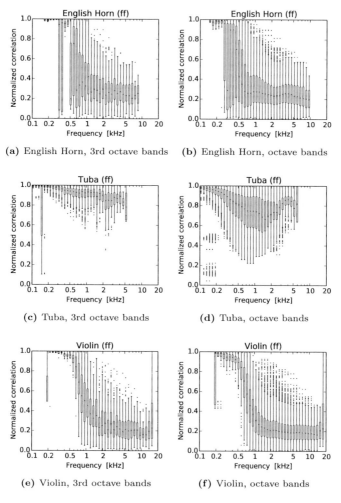

Figure 3.12.: Statistical analysis of correlation values for radiation patterns of woodwind, brass and string instruments in third octave band and octave bands.

scheme. Using a surrounding spherical microphone array for the measurement the angular resolution of the array is usually rather limited, so an accurate centering of the sound source is beneficial in order to reduce these aliasing errors.

An adjustment according to the geometric position in the space domain has the potential to decrease the effect of spatial aliasing. This can be done by a modification of the recorded signal according to Eq. (2.44) before performing the SHT. The averaging approach using magnitude data as described in Sec. 3.2.2 behaves less sensitive to displacement, as translation causes mainly phase differences at the microphone positions that is not accounted for during averaging.

3.4.1. Centering by minimizing the weighted SWS

One approach to find the origin of the emitted wave is to employ nonlinear optimization that minimizes a cost function. The choice of this cost function is an active topic of current research. Deboy et al. [Deb10] present two optimization criteria for deriving optimal center points using measurement data of a surrounding spherical microphone array with 64 sensors. Ben Hagai et al. [BH11] give a number of objective functions as follows that show its minimum values at or close to the physical center of sound radiation:

- J_0 - power of the zero-order harmonic

$$J_0 = 1 - \frac{|c_{00}|^2}{L_2}, \tag{3.5}$$

where

$$L_2 = \|c\|_2^2 = \sum_{n=0}^{N} \sum_{m=-n}^{n} |c_{nm}|^2. \tag{3.6}$$

- J_1 - power ratio

$$J_1 = 1 - \frac{\sigma_{\overline{N}}^2}{L_2}, \tag{3.7}$$

where

$$\sigma_{\overline{N}}^2 = \sum_{n=0}^{\overline{N}} \sum_{m=-n}^{n} |c_{nm}|^2. \tag{3.8}$$

- J_2 - center of power

$$J_2 = \sum_{n=0}^{N} \sum_{m=-n}^{n} \frac{n\,|c_{nm}|^2}{L_2}. \tag{3.9}$$

61

- J_3 - center of magnitude

$$J_2 = \sum_{n=0}^{N} \sum_{m=-n}^{n} \frac{n\,|c_{nm}|}{L_1}, \tag{3.10}$$

where

$$L_1 = \|c\|_2 = \sum_{n=0}^{N} \sum_{m=-n}^{n} |c_{nm}|. \tag{3.11}$$

In a study with analytic sound sources and the recording of a trumpet in the spherical array with the geometry as defined in Appendix A.1.4 the J_2 measure shows the best performance, but exhibits a non-convex behavior at higher frequencies of approx. 1 kHz. The deviations of the array radii of 2.5 cm have been compensated for using the far-field approximation of Eq. (2.44), but have shown not to have significant impact on the results. For details on this type of analysis, refer to Ben Hagai et al. [BH11]. Centering algorithms that show a better convergence for higher frequencies are currently being investigated by Shabtai et al. [Sha14].

3.4.2. Centering by minimizing phase transitions

An alternative approach regards only the phase of the signals, as it contains the information of the traveled distance of an acoustic wave. A sound source described by a multitude of monopoles located infinitesimally close to the geometric center result in constant phase values over the measurement sphere (with possible phase transitions at the notches in the spatial domain, as occurs e.g. with a dipole characteristic). Any displacement from the center point modifies the phases occurring at the sensors depending on the shorter or larger traveling distances from the sound origin to the microphones.

This phase shift can be observed in the radiation patterns of musical instruments as recorded with the spherical microphone array. Due to their consistent radiation patterns (cf. Sec. 3.3) and simple geometry brass instruments are expected to be modeled efficiently as single point radiators. As an example the radiation pattern of a trumpet player has been studied, whose placement is depicted in perspective photos with a white cross marking the approximate center of the microphone array in Fig. 3.13. It can be observed that for the this example the horn of the trumpet (which is the assumed origin of sound radiation) is located slightly displaced from the geometric center. Performing the complex analysis of a single tone (cf. Sec. 3.2.4), a continuous phase change over the measurement sphere is visible. As an example the radiation of the standard pitch A at approx. 440 Hz is depicted in Sec. 3.14a. The data can be plotted as a complex balloon plot (cf. Appendix A.4.3) at the discrete microphone values (colored dots) and as continuous function that was obtained performing the SHT with a maximum order of $n_{\mathrm{max}} = 4$ (cf. Sec. 3.2.1). Performing a far-field compensation as described in Sec. 2.3.5 to the positions of the horn opening of the trumpet, the compensated balloon plot in Fig. 3.14b consists of a constant phase indicating that the center point for that instrument was found correctly [Pol12e].[15]

This observation can be used to define suitable objective functions and using the phase as optimization criterion. The SWS of the complex function shows increasing degradation from aliasing for larger displacements. As the displacement causes different traveling times of the acoustic wave to the spherical surface, mainly phase shifts of the directivity occur if the two radii r_0 and r_1 in Eq. (2.44) are of similar value. Using the large argument approximation for the magnitudes of the spherical Hankel function (c.f. Eq. (5.8)) the magnitude of the sound pressure values is scaled by the ratio of the two radii r_0 and r_1. For directivity pattern obtained in the far-field as described in Eq. (2.1) the change of the magnitude of the directivity function is almost marginal, so increased aliasing components are not to be expected for source translation.

One possibility is thus to define the SHT of the original function and the magnitude values of the function as

$$f_{nm} = \mathcal{S}\left\{f(\theta, \phi)\right\} \qquad \text{and} \qquad (3.12)$$

$$f_{nm,\mathrm{abs}} = \mathcal{S}\left\{|f(\theta, \phi)|\right\} \qquad (3.13)$$

and using the summed energetic differences of the magnitude values of their spatial domains as the objective function as

$$f_{\mathrm{obj}} = \oint_{S^2} \left[\left|\mathcal{S}^{-1}\left\{f_{nm,\mathrm{abs}}\right\}\right| - \left|\mathcal{S}^{-1}\left\{f_{nm}\right\}\right|\right]^2 \, \mathrm{d}\Omega \to \min. \qquad (3.14)$$

Minimizing the objective function using a nonlinear optimization algorithm can thus be regarded as finding a focal point that avoids aliasing. As sources usually possess a rather smooth directivity pattern for low frequencies, the optimization is performed sequentially with rising frequency. The initial guess for the solution is set to be the geometric center of the array, while each processing step used the

---

[15]The plotted data has been compensated by $(\Delta x, \Delta y, \Delta z) = (35\,\mathrm{cm}, -2\,\mathrm{cm}, -7\,\mathrm{cm})$. This displacement is the result of the nonlinear minimization of Eq. (3.14).

previous result as the new starting value for the next frequency the optimizer is running on.[16]

The results of this algorithm for the trumpet can be observed individually for each partial tone in Fig. 3.15. While the fundamental frequencies are all detected at a sensible position (upper row), the higher partials show deviations. Results for

Figure 3.13.: Measurement of trumpet directivity in the surrounding spherical microphone array. Approximative center of microphone array marked with a white cross [BH11; Pol12e].

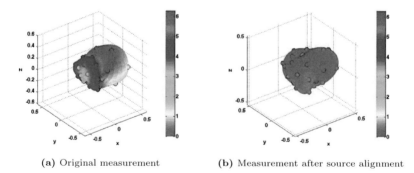

(a) Original measurement   (b) Measurement after source alignment

Figure 3.14.: Alignment of acoustical center using a far-field approximation, Trumpet playing the standard pitch A at 440 Hz, complex balloon plots [Pol12e]

---

[16]The *Matlab Optimization Toolbox* with the function `fmincon` has been used for the calculation.

frequencies below approx. 1.5 kHz are generally stable. Each tone of the trumpet converges at the location of approx. $\Delta x \approx 35$ cm, $\Delta y \approx -2$ cm and $\Delta z \approx -7$ cm, yielding to the compensated directivity pattern as depicted in Fig. 3.14b. For higher frequencies outliers occur, as the objective function does not possess a defined and unique minimum.

3.5. Summary natural sound sources

The sound radiation of natural sound sources varies, depending on their principle of tone generation. Computing the cross-correlation coefficients for all directivity patterns a musical instrument can excite allows to analyze the occurrences of similar or dissimilar pairs of directivity patterns in the same frequency range. The knowledge of these similarities yield guidelines for conservative simplifications of directivity patterns in order to derive a generic, frequency dependent directivity pattern that can be applied as a frequency dependent filter.

Sound sources that do not change their geometry significantly show mostly uniform behavior for wavelengths larger than the instrument size. Brass instruments, e.g., show to be represented well by a generic directivity pattern obtained by averaging in third octave bands. Woodwind instruments show frequency blocks of high correlation in low frequencies and erratic directivity patterns at higher frequencies. String instruments behave erratic for wavelengths smaller than approx. the instrument body. For larger wavelengths the directivity patterns are rather uniform with high correlation values.

Applying a translation of the recorded sources in post-processing, yields a different representations in terms of their spherical wave spectra. The directivity can be represented more efficiently for specific choices of the focal point used in vicinity of the physical origin of sound radiation. Acoustic centering of natural sources is a current topic of research and has been touched briefly in this chapter [Deb10; Sha14].

The obtained data can be used for the purpose of realistic auralization using the directional dependent sound radiation of musical instruments.

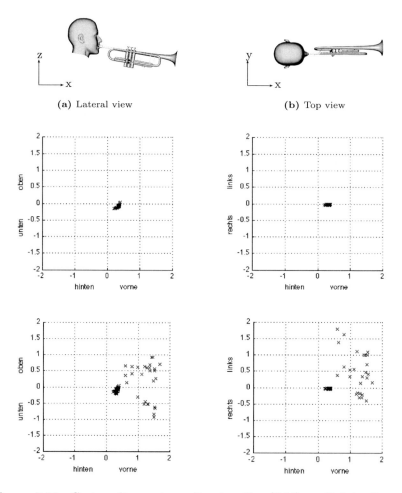

(a) Lateral view

(b) Top view

Figure 3.15.: Center alignment result using Eq. (3.14) applied to the extracted complex peak information of a trumpet. Upper graph shows the fundamental frequencies of all played tones ($f = 185\,\text{Hz}\dots 1.4\,\text{kHz}$), the lower graph shows the fundamental plus the first two higher harmonics ($f = 185\,\text{Hz}\dots 4.2\,\text{kHz}$).

4

Directivity patterns of binaural sound receivers

In this chapter the directivity at the receiver side of the acoustical transfer path is discussed. For a human listener the directivity pattern of the upper body with the head, torso and ear can be considered as the perceptually most important part in a binaural auralization of a room acoustical scene. Humans are capable of localizing sound sources with a remarkable accuracy, in particular using the direction of sound incidence and, to a smaller extent, the distance of the source to the listener [Bla97].

The directivity pattern of a human listener is described by the head-related transfer functions (HRTFs) usually obtained for sound incidence from a set of points located on a spherical surface around the listener's head and can be represented as spherical wave spectrum (SWS) using algorithms of Fourier acoustics. In the spherical harmonic domain the correct interpolation and extrapolation of HRTFs becomes feasible, as long as the spatial sampling is sufficiently dense in order to avoid spatial aliasing. The fruitful combination of binaural technology with Fourier acoustics allows algorithmic solutions for practical problems that have been solved previously with the effort of lengthy measurement sessions.

The HRTF datasets obtained in this chapter can be used for both measurement and simulation of room acoustics. Recent advances in the implementation of fast measurement methods for individual HRTFs are shortly reviewed (cf. [Mas12] and [Die12b]). It is shown that a variation of the focal point used for the decomposition in the Fourier domain can prove beneficial for a compact description of HRTFs as SWS.

Parts of this chapter have been published in [Pol11a], [Pol12c] and [Pol12a].

4.1. Obtaining head-related transfer functions

Head-related transfer functions (HRTFs) describe the transfer paths from acoustic sources located at arbitrary positions to the eardrum or the opening of the ear canal of a human listener. These functions can be utilized to calculate the signals arriving at the listener's ears originating from sound sources located at any point in space by applying convolution.

Blauert [Bla97] gives different definitions of the HRTFs, namely the free-field HRTFs (referenced to the sound pressure encountered in absence of the listener), the monaural HRTFs (referenced to the level in a specific direction) and the interaural HRTFs (the fraction between the sound pressure at both ears). The definition of the frequency dependent free-field HRTFs is given as

$$H^{[L/R]}_{\text{free-field}}(\theta, \phi) = \frac{H^{[L/R]}(r, \theta, \phi)}{H_0(r)} \tag{4.1}$$

with $H^{[L/R]}(\cdot)$ being the transfer paths from the source position at (r, θ, ϕ) to the left (L) and right (R) ear and $H_0(\cdot)$ being the transfer path of the sound source to an omnidirectional microphone mounted at the center point of the listener's head. The monaural HRTFs

$$H^{[L/R]}_{\text{monaural}}(\theta, \phi) = \frac{H^{[L/R]}(r, \theta, \phi)}{H(r, 90°, 0°)} \tag{4.2}$$

are referenced to the sound incidence from frontal direction.

The definitions of the HRTFs are valid for sources located in the far-field of the listener. For the measurement of HRTFs this implies that the distance between source and listener is supposed to be sufficiently large in order to avoid any change in the HRTFs with respect to distance. As the HRTFs are defined as referenced transfer functions, any coloration caused by the equipment, such as the sound source or the receiver (when using identical sensors) cancels out.

Comparing the definition of the monaural HRTFs with the directivity factor defined in Eq. (2.1) shows their equivalence. The set of HRTFs can thus be regarded as two channel directivity pattern, quantifying the sound incidence for left and right ear. As a consequence all the computations in the Fourier domain as formulated in Chapter 2 can be applied to HRTFs, opening up the field of Fourier acoustics to binaural audio processing.

Reciprocity states that physical transfer paths can be calculated in reversed direction (cf. Sec. 2.6.2). Applying reciprocity to HRTFs allows to consider them as solution to an exterior problem as described in Sec. 2.3.4. These exterior problems are more straightforward to solve than assuming sources in the room combined with the secondary sources at the human head and torso that also contribute to the resulting sound pressure at the ear openings.

4.1.1. Simulation of HRTFs

HRTFs can be derived using numerical simulations that calculate the scattering of sound at the human body. The geometry of the part of the body causing the most significant contributions is represented using a discrete mesh of the boundary surface of ear, head and torso. Such a mesh has to be sufficiently fine (as a rule of thumb a minimum of 6 nodes per wavelength are required). The simulation of high frequency solutions thus demands a large memory for the computation. In Fig. 4.1 the meshes used for the simulation of two different dummy-heads are depicted.

(a) HEAD acoustics HSU III mannequin

(b) Custom-made mannequin produced at ITA, RWTH Aachen University [Sch93]

Figure 4.1.: Artificial heads used for measurement and simulation: Meshes as used in the numerical model are depicted

Applying reciprocity the incoming wave arriving at the listener from different source positions located on a spherical surface can be considered as an outgoing radiation problem that can be solved analytical or numerical with standardized methods. Kahana et al. [Kah98] suggest to use reciprocity for speeding up the simulation of HRTFs using the *Boundary Element Method* (BEM). Various authors adopted this method for the efficient simulation of HRTFs [Kat01; Gum02b; Ota03; Fel08].

Deriving time domain data from frequency domain simulation

The numerical simulation of HRTFs can be performed either using time domain or frequency domain methods. While the former yields the head-related impulse response (HRIR) that can be used directly for further processing, frequency domain methods are more commonly found, yielding results for discrete frequencies. Due to the lack of DC component in the simulation and an upper band limit that is commonly employed it is usually not possible to directly apply the Fourier transform as described in Sec. 2.2 to obtain the related HRIR that is used for the convolution of signals.

Using prior knowledge of the impulse responses, allows to obtain sensible solutions for the simulated HRTFs that match the simulation results at the given frequency points and continues the spectrum for frequencies below and above the highest simulated frequency. This can be done adapting the methods of compressed sensing in order to derive the missing information by exploiting the additional knowledge of sparsity in time domain [Can06].

The first step is to choose the sampling rate and the impulse response length that fulfills the following criteria: The sampling rate is chosen high enough for sufficiently broadband HRTFs, while the length of the HRIR is set to resemble all significant binaural features. As the information content of an HRTF is included within approx. 1 m traveling distance of the acoustic waves, at least 3 ms are required for the length of the HRIRs [Len07].

The frequency spacing as used in the simulation and the impulse response length is matched in order to gain a representation for all frequency points after performing the DFT. A transformation matrix is derived for the chosen parameters, which represents the DFT by a matrix multiplication.

Using standard methods of compressed sensing allows to derive meaningful results, while block sparsity approaches that search for blocks of zeros are expected to perform even better for that particular problem. Using ℓ_1 minimization as suggested by Candès et al. [Can06] it is possible to extend the frequency range of the simulated data to DC and higher frequencies. This results in a HRIR whose spectrum matches the set of simulated frequencies and has the additional property of being compact in time domain. Applying conventional resampling, allows to transform the chosen sampling rate to a sampling rate supported by the used audio equipment.

Informal listening tests have shown good performance of the sparse time domain approach. More elaborate methods (such as the mentioned block sparsity approach) are expected to create more accurate time domain representations of simulated data obtained only at a limited number of frequencies.

4.1.2. Sequential measurements of HRTFs

HRTFs can be obtained using sequential measurements, employing a single loudspeaker in order to measure the transfer paths between the source and two microphones located at the ear openings. As the measurement procedure is very time consuming for high resolution data, usually HRTFs of artificial heads are measured using this technique. For human listeners the fast measurement methods as described in the following section are more suitable.

An example for a measurement setup designed for sequential measurements can be seen in Fig. 4.2. The artificial head is mounted on a continuously adjustable turntable for the horizontal orientation of the head. The trellis arm allows to modify the elevation angle in ranges between $0°$ and approx. $120°$ without distorting the sound field significantly. In order to obtain a full spherical measurement the dummy head can be mounted upside down and the measurement can be repeated in order to obtain data for the lower hemisphere of the HRTFs. This measurement setup requires the application of a suitable time window to eliminate the ground reflection that is always present in the hemi-anechoic chamber. Details of the described measurement setup can be found in Lentz [Len07] and Masiero [Mas12], cf. also Appendix A.1.1.

If performed in a quiet environment, this measurement setup provides for very precise results.[1] The uncertainties of the measurement results are caused mainly by a slight spatial deviation of artificial head position or receiver positions. As a result, the process of merging the data obtained at the upper and lower hemisphere can be demanding. Slight variations of the vertical positioning of the head can cause a discontinuity at the plane used for concatenation (usually located the at equator of the full sphere).[2] This can render further signal processing

---

[1] Measurement results of highest precision are obtained in during night time or weekends, when the noise from slamming doors and heavy machines close to the measurement room are absent.

[2] An example of such an discontinuity can be found in the measured HRTFs used for directivity synthesis on page 113.

71

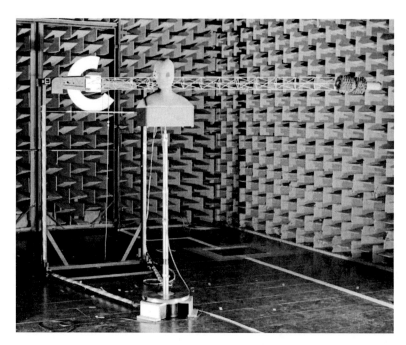

Figure 4.2.: Sequential measurement system to obtain HRTFs using high precision measurements of hemispherical data. For measurement data
covering the full sphere the artificial head can be mounted upside-
down in order to measure the lower hemisphere.

in the spherical harmonic domain difficult. Generally, mathematical algorithms
such as order dependent Tikhonov regularization when performing the SHT allow to minimize the influence of this discontinuous transition. Nevertheless, it
is advantageous to perform the acoustical measurement as precise as possible in
order to obtain the best possible representation of the HRTFs.

4.1.3. Fast HRTF measurement methods

Sequential measurement methods are usually too slow for the measurement of
HRTF for individual subjects. Several approaches can be found in literature to
measure the HRTFs of individual subjects in a fast manner, such as [Bro95] and
[Møl95]. Zotkin et al. [Zot06] use reciprocity for a fast measurement method of
individual HRTFs. Majdak et al. [Maj07] use specialized measurement signals to
accelerate the measurement of multiple transfer paths. Antweiler et al. [Ant09]

describe an accelerated measurement method of individual HRTFs using adaptive filtering to obtain a quasi-continuous representation in azimuthal direction.

Measurement Setup

In order to obtain high-resolution HRTFs in a comparative short measurement duration, Masiero [Mas12] developed a system that allows the rapid measurement of high-resolution HRTFs, which is depicted in Fig. 4.3. A set of 3840 measurement directions (40 elevations combined with 96 azimuthal positions, located approximately on a truncated Gaussian quadrature sampling scheme) is measured at a radial distance of $r_{LS} \approx 1 \, \text{m}$ [Pol12b]. The fast measurement duration is accomplished using a modified version of the *Multiple Exponential Sweep Method* (MESM) as mentioned in Sec. 2.6.1. The advantage of this method is the greatly reduced measurement duration of approx. six to seven minutes [Mas12; Pol12b].

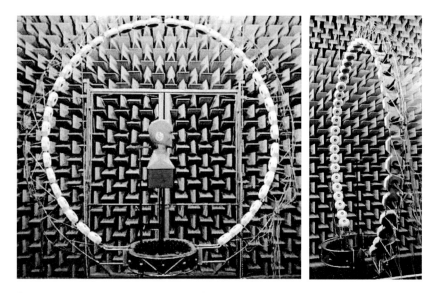

Figure 4.3.: Measurement system for the fast acquisition of HRTFs. Subject is rotated around the vertical axis (full rotation) while the arc of loudspeakers stays in fixed position.

Directivity pattern of loudspeakers

For high quality measurement results of individual HRTFs it is important that all significantly contributing scattering regions of the human body receive an acoustical wave of identical intensity. For that reason the loudspeakers used in the array should possess a smooth directivity pattern in frontal directions. The required opening angle of uniform sound distribution can be calculated geometrically by the equation

$$\theta_0 = \arctan\left(\frac{r_{\text{sources}}}{r_{\text{LS}}}\right) \tag{4.3}$$

relating the radius r_{sources} of a sphere around the head containing all significant scattering sources with the radial distance r_{LS} of the loudspeakers to the geometric center of the setup. For the HRTF measurement arc developed by Masiero [Mas12] this computes to an angle of $\theta_0 = 14°$ from the on-axis response of the loudspeaker which corresponds to an offset of $r_{\text{sources}} = 25\,\text{cm}$ from the geometric center. This area of contributing secondary sources is expected to be sufficient for precise measurement results of HRTFs. In Fig. 4.4 a schematic representation of the required area of equal sound incidence is depicted.

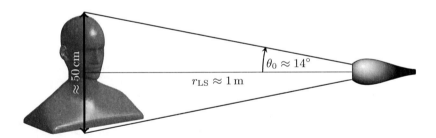

Figure 4.4.: Estimation of the required angle of even sound radiation for HRTF measurements, adapted from [Zil14]

A description of the optimization of the loudspeaker directivity patterns for this measurement setup is given in Appendix A.2.2.

Detection of loudspeakers positions

Another concern for accurate HRTF measurements are errors in loudspeaker positioning. These deviations can complicate a spatially continuous time-of-arrival detection as described in Majdak et al. [Maj13a] and Ziegelwanger et al. [Zie14], as they distort the measurement results, for radial deviations especially in terms of the arrival time of the acoustical wave.

The actual source positions of the waves emitted from the loudspeakers can be found using a set of receiver points on a microphone array mounted in-place of the listener in the center of the array. The arrival time of the impulse is calculated using a convolution with the minimum-phase version of itself and calculating the zero-crossings of the Hilbert transform of the result [Mas12].

A nonlinear optimization approach yields the positions of all used loudspeakers as a frequency independent position of the excited waves [Kre12; Mas12; Zil14; Pol14a]. The loudspeaker position with respect to the rotational axis of the turntable, can be stated as

$$\mathbf{x}_{\text{LS,opt},j} = \arg\min_{\mathbf{x}_{\text{LS},j}} \sqrt{\sum_{i,k} \left(\|\mathbf{x}_{\text{LS},j} - \mathbf{x}_{\text{mic},ik}\| - \tau_{\text{ijk}} \cdot c\right)^2} \qquad (4.4)$$

with j being the number of single loudspeakers, k being the number of measurement microphone in the array and i being the number of a set of sequential measurements using the microphone array within the loudspeaker array.

4.2. Low-frequency extension of measured HRTFs

Using measured HRTFs for means of auralization, it is important to cover the full audible frequency range. Measured data, however, usually suffer from insufficient signal-to-noise ratio (SNR) for low frequencies, due to the comparatively small size of the used loudspeakers in common measurement setups. These small speakers do not interfere as strong with the created sound field during the measurement and create a smoother directivity pattern in frontal direction due to their smaller aperture size (that evenly excite a larger range of SH orders, cf. Sec. 5.2). The excitation of low frequencies, however, requires large volume velocities to gain significant levels. According to Masiero [Mas12] the individual variation for HRTFs is very small for frequencies below 300 Hz, so all meaningful features of an individual head and pinna geometry are included in the frequency range of the measurement.

In order to restore the information in the low frequency range, the measurement result can be combined with alternative methods, such as numerical methods i.e. the *Boundary Element Method* (BEM), the *Finite Element Method* (FEM)

75

or the *Fast Multipole Method* (FMM). Zotkin et al. [Zot06] use a reciprocal HRTF measurement setup that is insufficient at frequencies below approx. 1 kHz and use the low frequency behavior of a set of intersecting sphere (also named descriptively as the *snowman model*) in order to gain a solution for the full audible spectrum.

High precision measurements of the ITA prototype dummy-head [Sch93] using the measurement system as described in Sec. 4.1.2 lack of sufficient SNR in the frequency range below approx. 200 Hz. These measured HRTFs have been augmented in this work by morphing the spatial characteristic of their responses for low frequencies to a numerical simulation of an artificial head. Using these simulation data only for the low frequency extension a much coarser mesh can be used resulting in significantly shorter calculation times compared to full frequency range simulations.

In terms of the representation of HRTFs as SWS, the kr-limit as frequency dependent order limit as described in Sec. 2.3.8 proves to be useful and allows to enhance the signal for order limited functions in the low frequency range. Moving asymptotically towards the DC-component, the set of HRTFs can be limited with a smaller and smaller specific order of SH coefficients without loosing accuracy. The DC-component itself can be calculated with a zero order limit yielding a constant value for all directions.

4.3. Range extrapolation of HRTFs

While the original definition of the HRTFs given by Blauert [Bla74] is limited to the auralization of sources in the far-field of the listener, attempts have been made to auralize sources in closer vicinity. Duda et al. [Dud98] study the effect of sources close to a spherical scatterer and conclude that the changes in the *interaural level differences* (ILD) are significant for closer distances than five times the radius of spherical head model, while the *interaural time differences* (ITD) are not significantly altered by close sources. In 2007 Lentz [Len07] evaluated the audibility of near sources recorded with an artificial head. He concludes that these effects become noticeable at source distances closer than 1.5 m and can be considered significant in ranges closer than approx. 60 cm with a higher sensitivity to lateral sources. In order to include these effects in real-time auralization systems, the HRTFs measured at different ranges (radial distances) are used in the virtual reality auralization of the virtual reality system at RWTH Aachen University [Len07].

Representing the HRTFs by their spherical wave spectrum (SWS), the radial filter for a (reciprocal) exterior boundary problem as described in Sec. 2.3.4 can be applied in order to compute the HRTFs for ranges that differ from the distance of the original data. The obtained transfer functions given as SWS can be converted to different ranges, allowing to auralize sound sources at different distances from the listener even if the set of original HRTFs is available only at one specific range [Dur04]. As this approach also provides (for a sufficiently dense spatial sampling) a physically correct interpolation of the angles of sound incidence, the HRTFs for any point in the three dimensional space can be computed.

In order to evaluate the theoretical formulation of the range extrapolation of HRTFs from Duraiswami et al. [Dur04], the algorithm is tested with both simulated and measured HRTFs. The cross-correlation of spherical functions as defined in Sec. 2.3.7 is used as a measure of the quality of extrapolation, regarding both inward and outward extrapolation.[3]

4.3.1. Principle of calculation

Applying Eq. (2.35) the SWS of the HRTFs can be obtained as

$$p_{nm}(r) = \mathcal{S}\left\{p(r, \theta_i, \phi_i)\right\}. \tag{4.5}$$

A frequency dependent order limit as described in Sec. 2.3.8 and suggested in the original article on range extrapolation of Duraiswami et al. [Dur04] is employed. The radius $r_{\min}$ encompassing all contributing sources is chosen to be 30 cm, equaling roughly the radius of the outermost edge of the dummy heads as depicted in Fig. 4.1.

The extrapolation of the HRTFs can then be calculated by applying Eq. (2.40) to the SWS using an order and frequency dependent ratio of outgoing spherical Hankel functions as

$$p_{nm}(r_1, k) = p_{nm}(r_0, k)\frac{h_n(kr_1)}{h_n(kr_0)}. \tag{4.6}$$

Inserting this result into Eq. (2.36) allows to derive spatially continuous function values for the HRTFs at any desired point in the 3D space outside a sphere

---

[3]The data used for the extrapolation were a subset of the the $1°/1°$ dataset in $5°/5°$ steps as measured by Lentz in 2001 and the set of HRTFs of different distances measured by Lentz in 2004, see Appendix A.3.2.

located at the center of the head with a radius $r_{\min}$ that contains all secondary scattering sources.

The normalized correlation coefficient in the spherical harmonic domain as defined in Eq. (2.59) provides a physically motivated measure for the similarity of the spherical shape of two functions. Although not explicitly measuring the perceptional impact, a mathematically high correlation of the functions also suggests high similarity in terms of human sound perception.

4.3.2. Results for measured HRTFs

To evaluate the performance of the range extrapolation on realistic data, the HRTFs at different ranges as measured by Lentz [Len07] are used to analyze the performance of the algorithm. The artificial head whose mesh is depicted in Fig. 4.1b has been used for the study.

The magnitude values of the correlation between measured HRTFs and extrapolated functions are plotted in Fig. 4.5 over frequency. The solid (red) line describes the quality of outward extrapolation, the dashed (gray) line shows the quality of inward extrapolation. The dotted (blue) line acts as reference and shows the cross-correlation coefficients of the original functions as measured in two distances. It can be seen that the extrapolation between the distances 30 cm and 2 m does not increase the cross-correlation values of the original function at a specific range and the extrapolated result for the same range. According to the measurement report the data used in Fig. 4.5a have been recorded in different sessions, so additional uncertainties from the modified measurement setup can be expected.[4]

A comparison of two functions recorded with the same measurement setup is depicted in Fig. 4.5b. Here a slight improvement can be noticed for the extrapolations both in inward and outward direction in comparison to the cross-correlation values of the original functions. The SHT was performed using order dependent Tikhonov regularization as described in Sec. 2.4.4, but the correlation plots for inversion by pseudoinverse show only very minor differences.

The obtained frequency dependent correlation coefficients of measurement result and range extrapolation from data measured on a different radial distance are surprisingly low. According to this measure the range extrapolation shows only

---

[4]The arm and turntable measurement system as described in Appendix A.1.1 has been used by Lentz [Len07], using different arm constructions for the measurements at different ranges.

marginal improvement or even degradation with respect to the HRTFs obtained at another radial distance.

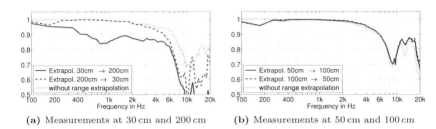

(a) Measurements at 30 cm and 200 cm   (b) Measurements at 50 cm and 100 cm

Figure 4.5.: Correlation coefficients of range extrapolation results and measured HRTF data using order dependent regularization, cf. [Pol12a].

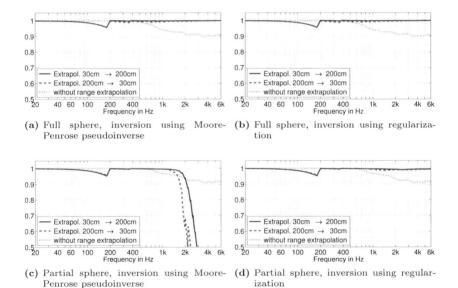

(a) Full sphere, inversion using Moore-Penrose pseudoinverse

(b) Full sphere, inversion using regularization

(c) Partial sphere, inversion using Moore-Penrose pseudoinverse

(d) Partial sphere, inversion using regularization

Figure 4.6.: Correlation for range extrapolation results calculated from data covering a full sphere (top) and a partial sphere ($\theta \leq 120°$) using Moore-Penrose pseudo-inverse and Tikhonov regularization with $\varepsilon = 10^{-4}$ (bottom), cf. [Pol12a].

4.3.3. Results for simulated HRTFs

In order to evaluate the algorithm on precise data, a set of HRTFs has been obtained using the *Boundary Element Method* (BEM) of *LMS Virtual.Lab*. The results for the inversion using order dependent Tikhonov regularization and using pseudoinverse are depicted in the upper row of Fig. 4.6. The simulation was performed at radial distances of 30 cm and 2 m from the center of the head (point in-between the ear openings).

Apart from the dips in the low frequency region (e.g. at around 180 Hz) caused by the frequency dependent order truncation (cf. Sec. 2.3.8), the range extrapolation yields almost perfect results with a correlation coefficient of nearly unity for almost all frequencies. A relaxation of this order limit is expected to enhance the result in this low frequency range [MT11]. In Fig. 4.7 the magnitude levels of the horizontal direction of the HRTFs are plotted over frequency, in the top row the BEM simulation results for 30 cm and 2 m of the artificial head as depicted in Fig. 4.1b and in the bottom row the extrapolation results for inward extrapolation to 30 cm and the outward extrapolation to 2 m.

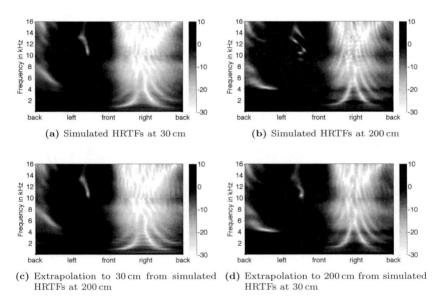

(a) Simulated HRTFs at 30 cm (b) Simulated HRTFs at 200 cm

(c) Extrapolation to 30 cm from simulated HRTFs at 200 cm (d) Extrapolation to 200 cm from simulated HRTFs at 30 cm

Figure 4.7.: Simulated azimuthal HRTFs at different distances (top) and range extrapolation results (bottom) [Pol12a].

It can be seen that the range extrapolation yields visually a good match of original and the appropriate extrapolation, independent of the direction of extrapolation. The fine structure of the occurring peaks and notches agree with high precision. The discontinuity at 10 kHz is caused by a change of the mesh resolution at that frequency.

4.3.4. Conclusions range extrapolation of HRTFs

The performance of the range extrapolation algorithm seems highly dependent on the quality of the used input data. While simulated data can be extrapolated perfectly, the studied measurement data show rather mediocre results. While data obtained at a single measurement session show a slight improvement from range extrapolation, data measured using different setups could not be enhanced by range extrapolation methods using the cross-correlation as measure for similarity. In a joint publication of the author with the audio group of IRCAM in Paris another set of measurements is discussed, showing sightly better results for different measurement data, cf. to Pollow et al. [Pol12a] for further details.

4.4. Reconstruction of missing data of HRTFs

In real measurement scenarios for HRTFs often the data at lower elevation angles cannot be obtained, as e.g. using the fast measurement system described in Sec. 4.1.3. In order to study the effect of having only a subset of the data available, BEM simulation data has been truncated at elevations below $150°$, which represents the identical part of the sphere as in the measurement setup.

The HRTFs available only on partial spheres were used for the ISHT employed in the range extrapolation using inversion by pseudoinverse and regularized inverse. In the second row of Fig. 4.6 the extrapolation results for such truncated datasets are depicted. Using a regularization approach the high performance as with the full dataset can be obtained, at least up to 6 kHz, which was the upper frequency limit for the simulated data.

In a second example, the measurement data of an artificial head in the measurement arc as depicted in Fig. 4.3 is used. This setup again covers elevations from the north pole to $\theta \approx 150°$, leaving an area without data at the southern part of the sphere. In Fig. 4.8a the measured HRTFs are depicted in two view angles as interpolated magnitude values plotted on the sphere for frequencies of 500 Hz, 1 kHz, 5 kHz and 15 kHz. The positions of the loudspeakers were detected by

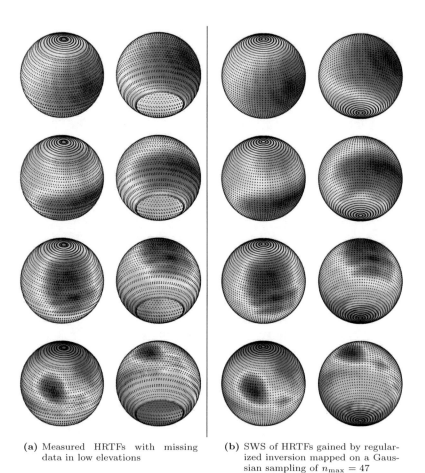

(a) Measured HRTFs with missing data in low elevations

(b) SWS of HRTFs gained by regularized inversion mapped on a Gaussian sampling of $n_{\max} = 47$

Figure 4.8.: Magnitude of HRTFs from two view angles at 500 Hz, 1 kHz, 5 kHz and 15 kHz (from top to bottom) for measured data and the derived SWS. The black dots mark the sampling points of the discrete spherical functions.

applying a non-linear optimization of the positions as formulated in Eq. (4.4). A phase compensation for the slight deviations in traveling time of the acoustic wave (in the range of a few centimeters) has been applied prior to processing. The order dependent Tikhonov regularization as given in Sec. 2.4.4 is used for the SHT as

$$\hat{\mathbf{p}} = \left(\mathbf{Y}^H\mathbf{Y} + \varepsilon\mathbf{D}\right)^{-1}\mathbf{Y}^H\mathbf{p} \qquad (4.7)$$

with the diagonal matrix $\mathbf{D}$ as defined in Eq. (2.85) and the regularization constant of $\varepsilon = 10^{-5}$.

Having found the spherical wave spectrum of the HRTFs it is feasible to reconstruct the spherical function for any arbitrary direction. An irregular distribution of sampling points that possibly does not cover the full sphere, as is often found in practical measurement setups, can thus be used to derive a representation in the spherical Fourier domain, as long as the sampling has been sufficiently dense not to create aliasing components (cf. Sec. 2.5). An example can be seen in Fig. 4.8b where measured data of the setup as described in Sec. 4.1.3 have been processed. The target sampling scheme has been set to a Gaussian sampling of a maximum order of $n_{\mathrm{max}} = 47$.

4.5. Representation of HRTFs with varying focal point

HRTFs are usually measured or simulated at equidistant points from the center of the listeners head. The opening of the ears are thus not located in the geometric center of the sampling points. From research on loudspeakers it is known that the acoustic center is generally frequency dependent and can even be located on the outside of the device [Van06].

In order to apply this idea to HRTFs Richter et al. [Ric14] evaluate the acoustic center of HRTF datasets using the objective function J_2 as defined in Sec. 3.4.1 and employing nonlinear optimization. Both measured and simulated HRTF datasets suggest to have the acoustic center of their reciprocal radiation for low frequencies at or slightly beyond the ear opening (outside of the head), similar to the known results from loudspeakers.

Using the found center of the spherical wave as new choice of focal point for the SHT, accurate representation of HRTFs with a lower order limit can be obtained. In Fig. 4.9 the required order for the match of a specific fraction of the signal energy (squared magnitudes) is plotted over frequency for a BEM simulated HRTF dataset. Especially for coarser approximations, the required number of

SH coefficients can be reduced significantly. An energetic match of 95% can be achieved at approx. 7 kHz using a maximum order of 4 for the ear-centered SHT, while order 10 is required for the head-centered representation. This order reduction allows significant speed-ups for real-time applications [Ric14]. For very high precisions, however, the required SH order is not diminished as can be seen for the dash-dot line.

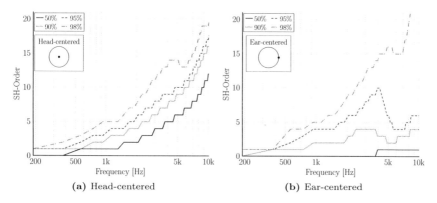

(a) Head-centered   **(b)** Ear-centered

Figure 4.9.: Maximal spherical harmonic order needed to represent BEM simulated HRTFs as SWS with a given accuracy expressed as a fraction of the signal energy [Ric14]

4.6. Summary binaural sound receivers

A human listener receives sound with a specific coloration depending on the angle of sound incidence. The influence on the spectrum and the time structure of the received sound can be represented by the head-related transfer functions (HRTFs), obtained by measurements or simulation methods.

These directional dependent transfer functions can be considered as complex valued directivity patterns of the human listener with two channels, one for each ear. The HRTFs can thus be processed using the analytic methods of Fourier acoustics as described in Chapter 2, allowing them to be included in the given mathematical framework used for the directivity synthesis in rooms.

The representation of the HRTFs as spherical wave spectra (SWS) furthermore allows to apply sensible post-processing methods in order to e.g. compensate insufficient dynamics of measurement results in the low frequency range or incomplete measurement data that does not cover all possible angles of sound incidence.

In the Fourier domain the methods of acoustical holography in spherical coordinates can be employed, making it possible to calculate near-field effects of HRTFs. It has been shown that data of high quality is essential for precise extrapolation results. While numerical simulations show perfect results, HRTFs that have been measured in different setups and sessions could not prove the same findings obtained from the simulation data.

The choice of focal point for the representation of HRTFs as spherical wave spectra influences the occurring coefficients. Choosing the focal point close to the physical point of sound reception allows to represent the spherical function with a lower order limit compared to focal points farther away from the acoustical center. This is beneficial for avoiding aliasing artifacts for a limited spatial resolution as commonly encountered in practical setups.

5

Including directivity patterns in room acoustic measurements

In this chapter the implementation of arbitrary source directivity in room acoustical measurements is described. As the directivity patterns of the measurement sensors are inextricably linked to the measurement results, the respective patterns have to be in effect at the time of measurement, either directly or as superposition result composed from a suitable set of directivity patterns.

At the receiver side compact spherical microphone arrays are commonly used to implement variable directivity patterns,[1] at the sender side spherical loudspeaker arrays with multichannel excitation can be used. In this chapter the focus is set on the implementation of directivity patterns at the sender side, it order to facilitate the measurement of the full transfer path with flexible source and receiver directivity in combination with a spherical microphone array.

The international standard ISO 3382 for room acoustical measurement is reviewed and technical sound sources that can be used for the measurement with and without directivity synthesis are described. Using an analytic model for spherical sound sources allows rapid prototyping of specialized sources. Having constructed the device, the measured directivity pattern of the source can be used instead of the analytically modeled radiation, yielding more accurate results.

Finally, applications are presented that employ the excitation or superposition of arbitrary directivity patterns using multichannel sound sources. To verify the method, a technical sound source has been used as target source allowing to compare the measured room impulse response (RIR) with the superposition result of many RIR measurements of known directivity patterns.

Parts of this chapter have been published in [Pol07], [Pol09c], [Pol11b], [Pol12d] and [Pol13a].

---

[1] Commercial solutions for compact spherical microphone arrays exist, as well as algorithms for the conversion of the recorded sound fields to spherical wave spectra [Mey02; Raf05; Raf07; Li07].

5.1. Standardized room acoustic measurements

In room acoustic measurement tasks the directivity patterns of source and receiver are usually not taken into account and implicitly assumed to be of omnidirectional type. When performing measurements according to ISO 3382 [Iso], the
sound source used in the measurement has to have an omnidirectional directivity
pattern. The norm defines a maximum level variation over a set of horizontal
directions, as given in the following table:[2]

Frequency band	125 Hz	250 Hz	500 Hz	1 kHz	2 kHz	4 kHz
Allowed deviation[3]	±1 dB	±1 dB	±1 dB	±3 dB	±5 dB	±6 dB

Spherical loudspeakers arrays are commonly used due to their uniform behavior
for all possible angular directions. Placing a set of transducers that are distributed equally in a spherical chassis yields rather uniform directivity patterns
that fulfill the required omnidirectionality as demanded by ISO 3382. Commonly
the Platonic solids (a set of regular, convex polyhedrons) are used as geometry
of such devices, e.g. the loudspeakers in shape of a dodecahedron: At each of its
twelve faces a transducer of identical size is mounted, yielding a self-repeating
arrangement of loudspeakers [Pol09c].

In Fig. 5.1 the three-way system as used in the Institute of Technical Acoustics
of RWTH Aachen University is depicted, consisting of a 12-channel spherical
high frequency unit (used for frequencies above approx. 1.5 kHz), a 12-channel
mid-range unit (used for frequencies between 150 Hz and 1.5 kHz) and a single
channel subwoofer unit (used for frequencies below 150 Hz). Subdividing the complete spectrum in these frequency ranges allows to fulfill ISO 3382 well beyond
requirements on the source directivity pattern as given above.

5.2. Analytic model for spherical loudspeakers

The directivity pattern of a spherical sound source with a radially vibrating
circular area can be calculated analytically using the spherical cap model [Wil99;
Zot09b]. This model is useful for the simulation of spherical sound sources in
particular for the analysis of new shapes of spherical loudspeaker designs.

---

[2]These requirements are only demanded for the horizontal plane, yielding to some smart
loudspeaker designs that circumvent the norm by being only omnidirectional in the horizontal
plane.

[3]The given limits of allowed deviations have to be averaged over an angle of 30°.

Figure 5.1.: Dodecahedron loudspeakers as used for room acoustical measurements. Here a three-way system, consisting of two spherical dodecahedron loudspeakers of different sizes and a subwoofer unit is depicted.

5.2.1. Aperture function to model membrane vibration

The area of the vibrating membrane in an otherwise solid sphere is modeled as the *aperture function* $a(\theta, \phi)$ that can be defined as a continuous spherical function. This aperture function is defined as rotational symmetric function with respect to the z-axis as

$$a(\theta, \phi) = a(\theta) = 1 - \varepsilon \left(\theta - \frac{\alpha}{2} \right) \tag{5.1}$$

with α being the aperture angle and $\varepsilon(x)$ being the *Heaviside function* (or *unit step function*) [Pol09c]. The aperture function is visualized in Fig. 5.2, showing the vibrating area as dark spherical cap. The spherical wave spectrum (SWS) of this aperture function located at the northern pole can be calculated by a spherical harmonic transform (cf. Eq. (2.35)) as

$$a_{nm} = \mathcal{S}\left\{ a(\theta, \phi) \right\} = \delta_{m0} \cdot a_n = \delta_{m0} \cdot \sqrt{\pi(2n+1)} \int\limits_{\cos\frac{\alpha}{2}}^{1} P_n(x)\, \mathrm{d}x \tag{5.2}$$

where $P_n(x)$ is the Legendre polynomial of order n and δ_{m0} being the Kronecker delta function as defined in Eq. (2.33). The SWS of the rotationally symmetric aperture function has only non-zero values for degrees of zero ($m = 0$).

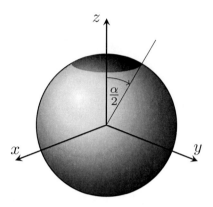

Figure 5.2.: Aperture function as used for the analytical spherical cap model. Dark section vibrates with a specific radial velocity, while the lighter gray section is assumed to be a solid hard sphere.

5.2.2. Total surface velocity and acoustic impedance

The aperture function given as SWS can be oriented to the locations of the membranes of the spherical loudspeaker array by implementing rotation using multiplication with the Wigner-D matrix and subsequent summation (cf. Eq. (2.50)) as

$$a_{nm,l} = \sum_{k=-n}^{n} a_{nk} \cdot D_{k,m}^{(n)}(0, \theta_l, \phi_l) \tag{5.3}$$

with (θ_l, ϕ_l) being the orientation of the membrane l of the dodecahedron loudspeaker.[3] The complete list of membrane angles is given in Appendix A.1.5. Alternatively the rotation can be implemented by a convolution on the sphere [Wil99; Pol09c].

The total surface velocity v_{nm} for a given set of cap velocities v_l can be given as

$$v_{nm} = \sum_{l=1}^{L} v_l \cdot a_{nm,l} \tag{5.4}$$

with L being the number of membranes of the spherical loudspeaker and $a_{nm,l}$ being the SWS of the aperture function of the lth membrane.

---

[3]The first rotation around the z-axis can be disregarded as the spherical cap model has a rotational symmetry.

The acoustic impedance of a spherically shaped body to the outside medium due to its radial vibration can be expressed as SWS as [Wil99]

$$Z_{nm} = \frac{p_{nm}(r_0)}{v_{nm}} = -j\rho_0 c \frac{h_n(kr_0)}{h_n'(kr_0)} \tag{5.5}$$

with r_0 being the radius of the spherical radiator and $h_n'(x)$ being the derivative of the spherical Hankel function. This equations states the connection between the surface velocity of the spherical body and the resulting sound pressure on the surface for any possible velocity distribution of the spherical surface.

5.2.3. Radiated sound pressure

The radiated sound pressure can be computed by applying the radial wave propagation for the exterior problem as formulated in Eq. (2.40) as

$$p_{nm}(r) = \frac{h_n(kr)}{h_n(kr_0)} \cdot p_{nm}(r_0). \tag{5.6}$$

Combining this equation with the impedance, the calculation of the surface velocities, and the definition of the aperture functions, the pressure at any point in space can be given analytically as

$$p_{nm}(r) = \frac{h_n(kr)}{h_n(kr_0)} \cdot Z_{nm} \cdot a_n \cdot \sum_{l=1}^{L} v_l \cdot D_{0,m}^{(n)}(0, \theta_l, \phi_l). \tag{5.7}$$

with $D_{0,m}^{(n)}(0, \theta_l, \phi_l)$ being the Wigner-D matrix that implements the rotation of the aperture function a_n to the positions of the L membranes with the complex, scalar velocities v_l.

The large argument condition

$$\lim_{kr \to \infty} h_n^{(2)}(kr) = j^{(n+1)} \frac{e^{-jkr}}{kr} \tag{5.8}$$

yields the far-field approximation of the sound pressure as used for the calculation of the directivity [Wil99].

From Eq. (5.7) it becomes obvious that the possible directivity patterns created by a given loudspeaker configuration depend directly on the aperture function a_n. Whereas the aperture functions differ when changing the orientation of the membrane, the order dependent signal energy stays constant for a fixed membrane size.[4] It is computed by the geometrically summed squared magnitudes of all coefficients of single SH order n as

$$E_{\mathrm{a}}(n) = \sum_{m=-n}^{n} |a_{nm,l}|^2 = |a_n|^2 \qquad (5.9)$$

with a_n being the SWS of the spherical cap aperture function. This value is constant for all possible membrane orientations l as it is invariant to rotation (cf. Eq. (2.51)). Using Parseval's identity as defined in Eq. (2.37) the limiting value for the infinite sum of the order dependent signal energy of the aperture function can be computed as

$$\sum_{n=0}^{\infty} E_{\mathrm{a}}(n) = \oint_{S^2} |a(\theta)|^2 \, \mathrm{d}\Omega = 2\pi \left[1 - \cos \frac{\alpha}{2} \right]. \qquad (5.10)$$

5.3. Design of compact spherical loudspeaker arrays

Using Eq. (5.7) the directivity patterns of a spherical sound source can be calculated analytically for a given set of membrane velocities. The order dependent signal energy of the aperture function as defined in Eq. (5.9) gives insight with respect to the feasible excitation patterns using a specific source geometry. The analytic model can be used for the analysis of existing devices or the rapid prototyping of specialized devices. Two examples are given in this section, extending an existing dodecahedron loudspeaker for multichannel excitation and the design of an optimized device for directivity synthesis using sequential measurements.

5.3.1. Dodecahedron loudspeakers as multichannel source

Dodecahedron loudspeakers used as single channel devices excite all membranes simultaneously with an identical signal. These devices are commonly employed for measurement tasks in room or building acoustics due to their omnidirectional radiation. Using a modification of the spherical loudspeaker as depicted in Fig. 5.1 (middle picture) each transducer can be excited with an individual signal. This 12 channel loudspeaker array can radiate with adjustable directivity depending on the applied signals. The inner volume is divided in 12 separate

---

[4]The definition of signal energy obtained from Parseval's identity in Sec. 2.3.3 is used.

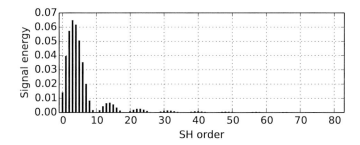

Figure 5.3.: Order dependent signal energy of the aperture functions of dodec-ahedron loudspeakers as depicted in Fig. 5.4.

chambers to provide individual air volumes for each transducer in order to min-imize cross-talk effects.

In Fig. 5.3 the signal energy of the aperture function for the mid-range dodec-ahedron loudspeaker[5] is plotted over the SH order n. At certain orders, e.g. at $n = 10$, the signal energy is negligibly small. Using a fixed geometry of mem-brane size and sphere radius, this energy distribution is unchanged, independent of tilt or rotation of the device, so the impact of the notches can only be reduced by changing the ratio of the sphere radius and the membrane size.

Using the multichannel loudspeaker for sequential measurements, a tilt of the device can be beneficial as it breaks the symmetry of its geometry. In the case of the dodecahedron loudspeaker, the device has been used sequentially in different positions rotated vertically around the z-axis. As the standard orientation of the dodecahedron depicted in Fig. 5.4a has a $120°$ rotational symmetry and only four unique elevations of the membranes, the tilted version in Fig. 5.4b has its membranes at 12 unique elevations and thus allows greater degree of freedom.[6] This tilted version is used for the synthesis of room impulse responses with a specific directivity pattern, as described in Sec. 5.4.2.

---

[5]The simulation was performed for an array radius of 15 cm and a membrane radius of 5.2 cm, which are the dimensions of the mid-range dodecahedron loudspeaker as built and used by the Institute of Technical Acoustics, RWTH Aachen University.

[6]In Appendix A.1.5 the membrane positions of the dodecahedron loudspeaker with and without tilt are listed.

(a) Dodecahedron loudspeaker mounted for directivity measurements

(b) Tilted dodecahedron loudspeaker mounted for room acoustic measurements [Kun11]

Figure 5.4.: 12 channel dodecahedron loudspeaker in normal position and tilted in order to maximize the variation in elevation.

5.3.2. Optimized sound source for broadband excitation

In order to develop an efficient measurement procedure, Klein [Kle12a] constructed an optimized measurement device that provides broadband excitation, both in terms of time-spectral frequency and spherical-wave-spectral frequency. Using a set of different transducer sizes allows to overcome the notches that occur when using only a single membrane size. The prototype for the optimized spherical sound source as depicted in Fig. 5.5a consists of a spherical chassis with a radius of $r = 20\,cm$ and is equipped with transducers of three different membrane sizes, using four 5-inch transducers, twelve 2-inch and twelve 3-inch transducers. The device contains a step motor mounted in the sphere and can be placed on a turntable as shown in Fig. 5.5b.

Choosing the geometry of the device

As commercially available transducers have standardized membrane sizes, the sphere radius has been chosen in a way to ensure broadband behavior in the SH domain. In Fig. 5.6 the signal energy of the aperture functions for conventional

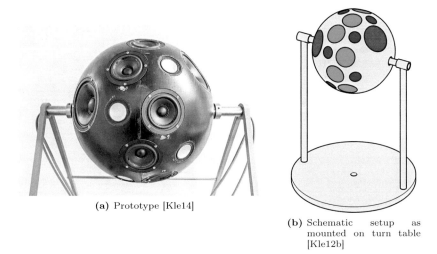

(a) Prototype [Kle14]

(b) Schematic setup as mounted on turn table [Kle12b]

Figure 5.5.: Optimized spherical loudspeaker array as measurement source for flexible directivity control in post-processing. The spherical chassis can rotate around the vertical and horizontal axis.

2-inch, 3-inch and 5-inch transducers mounted in a sphere with a radius of 20 cm are plotted over the spherical harmonic order n.

Plotting the logarithmic signal energy (normalized by membrane area) of the aperture functions of the three transducer types in the given sphere in Fig. 5.7 shows, that the criterion of broadband excitation up to a maximum order of approx. 60 is well fulfilled.

Transducer placement concept

The placement of the transducers has been arranged to place the elevation of the center points of the loudspeaker membranes at the elevations occurring in the Gaussian quadrature sampling (cf. Sec. 2.5.2), so that using a rotating turntable can yield full Gaussian sampling schemes with minor deviations (due to practical constraints in the mechanical construction). The device has been constructed for two possible spatial resolutions: covering a Gaussian sampling scheme of order 11 or order 23, by using 24 or 48 azimuthal orientations at one or two orientations around the horizontal axis, respectively. The latter uses for times as many positions and thus requires substantially longer measurement durations.

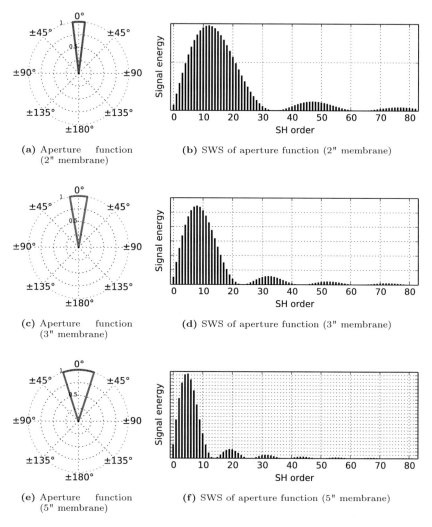

(a) Aperture function (2" membrane)

(b) SWS of aperture function (2" membrane)

(c) Aperture function (3" membrane)

(d) SWS of aperture function (3" membrane)

(e) Aperture function (5" membrane)

(f) SWS of aperture function (5" membrane)

Figure 5.6.: Signal energy of aperture function of 2", 3" and 5" transducers on a sphere of 20 cm radius. The distance of the horizontal grid lines corresponds to a fixed value for all plots.

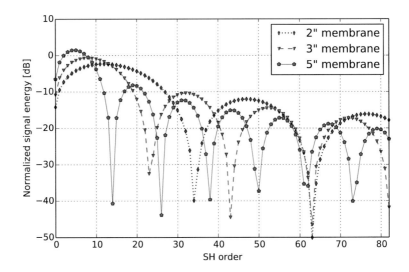

Figure 5.7.: Logarithmic signal energy of aperture functions 2", 3" and 5" transducers on a sphere with a radius of $r = 20\,\mathrm{cm}$ normalized by membrane area

Considering the acoustic impedance and the radial wave propagation term as used in the analytic model of the sound radiation in Sec. 5.2, it can be concluded that low frequencies only need to be excited up to a lower maximum order. A coarser sampling is thus sufficient for the largest membranes used in the lower frequency range and allows to make better use of the limited space on the spherical surface.

The list of transducer orientations in the prototype can be found in Appendix A.1.6, for further details on the transducer placement concept and measurement strategies refer to Klein [Kle12a].

5.3.3. Using measured directivity patterns for the synthesis

Having constructed the prototype the directivity patterns of the spherical loudspeaker array can be compared to results of the analytic model. As some differences occur, the measured patterns are used for further processing to obtain higher accuracy. These measured directivity patterns have been used e.g. for the

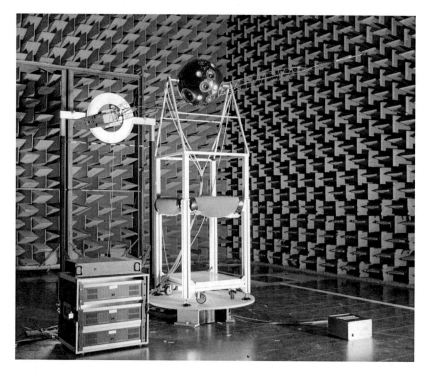

Figure 5.8.: Measurement setup for high-resolution directivity data of the optimized technical sound source.

synthesis of room impulse responses as described in Sec. 5.4.2 and the directivity synthesis as shown in Sec. 5.4.3.

For the measurement of technical sound sources, computerized adjustable positioning systems can be used as described in Sec. 4.1.2. Instead of a loudspeaker mounted on the trellis arm as used for the measurements of HRTFs, a microphone is fixed allowing flexible positioning at elevations up to approx. 120°. The directional dependent radiation of the measurement source can thus be obtained sequentially with very high precision. In Fig. 5.8 the measurement setup as used for technical sound sources is depicted.

The analytic cap model radiation and the measurement results of the optimized spherical loudspeaker are shown in Appendix A.3.3 for all three membrane types. At some frequencies deviations can be seen, while for most frequencies the analytic calculation and the measured directivity pattern show good agreement.

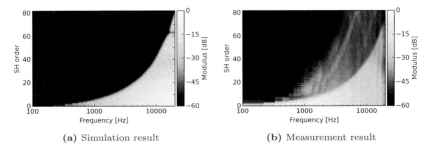

(a) Simulation result   (b) Measurement result

Figure 5.9.: Levels of the SWS of the sound pressure that can be radiated by
the optimized spherical loudspeaker array [Kle14]. The normalized
maximum values for all transducer types are depicted.

The plots in Fig. 5.9 quantify the maximum levels of the SWS of the sound
pressure radiated by the optimized spherical loudspeaker array with its three
different transducer types. As a rotation of the source around the z-axis allows
to distribute the energy in a specific SH order to different degrees (cf. Eq. (2.51)),
the values depicted over frequency are summed over all SH degrees and normalized to the global maximum, cf. [Kle14].[7]

In Fig. 5.9a the simulation result using the spherical cap model radiation as
defined in Eq. (5.7) is used, while in Fig. 5.9b the actually measured directivity
patterns are used for the calculation. The figures visualize the orders of SWS
(summed over all degrees) of the directivity patterns that can be excited over
frequency using a logarithmic color scale. For the simulation the expected notch
of excitation at orders of around 62 (that is also visible in Fig. 5.7) can be seen
in Fig. 5.9a.

The measurement results show excitation of higher orders as analytically predicted (cf. Sec. 2.3.8). An explanation is the influence of reflections occurring at
the outer frame structure of the physical setup of the loudspeaker array, yielding
a larger area of contributing secondary sources and consequently to higher orders
in the radiation.

If regarding aliasing components of 30 dB below the maximum levels as significant, it can be seen from Fig. 5.9b that the maximum order of the used Gaussian
sampling scheme is reached at a frequency of approx. 8 kHz, but only impacting

---

[7]To be able to cover all possible components by rotation, the directivity patterns of the
measurement device may not have a rotational symmetric pattern with respect to the z-axis
for any order.

the results for frequencies higher than 10 kHz because the aliasing is not affecting low orders immediately using Gaussian quadrature (cf. Sec. 2.5.2).

5.4. Applications of directivity synthesis

Spherical loudspeaker arrays can be used in various room acoustical applications in order to implement directivity patterns in room acoustic measurements. Some examples of directivity synthesis are given as follows.

In Sec. 5.4.1 the analytic cap model of the spherical loudspeaker array is used to synthesize averaged musical instrument directivity patterns that were obtained as described in Sec. 3.2.2.

Sec. 5.4.2 describes the use of the tilted dodecahedron loudspeaker mounted on a turntable for the synthesis of directivity patterns of higher orders using sequential measurements. The target directivity has been set as the directivity of a technical sound source. This allows to validate the method by comparing the RIR of the synthesis result and the RIR obtained by the measurement with the same technical sound source used as exciter.

Sec. 5.4.3 illustrates the synthesis result using measured directivity patterns of the optimized spherical sound source as described in Sec. 5.3.2. As target functions for the synthesis an infinitesimally narrow beam and a set of high-resolution HRTFs measured as described in Sec. 4.1.2 has been used. The influence of different regularization approaches on the resulting directivity pattern is analyzed and visualized.

5.4.1. Directivity synthesis of musical instruments radiation patterns: Simultaneous approach using a dodecahedron loudspeaker

In this section a multichannel dodecahedron loudspeaker is used for the synthesis of directivity patterns of musical instruments. These patterns have been obtained using an averaging approach (cf. Sec. 3.2.2) and are available as magnitude values p_i (without phase relations) recorded at 24 microphones (cf. Appendix A.1.3)[8] written as vector

$$\mathbf{p}_{\mathrm{mic}} = \mathrm{vec}\{p_i\}. \tag{5.11}$$

---

[8]The directivity used in the plots has been derived from recordings of longer passages of music using averaging and interpolation of the magnitudes to obtain a generic directivity over frequency [Len07].

A matrix $\mathbf{M}$ can be defined from Eq. (5.7) after spatial sampling using Eq. (2.56), mapping each membrane velocity vector

$$\mathbf{v} = \text{vec}\{v_l\} \tag{5.12}$$

to the complex sound pressure values

$$\mathbf{p}_{\text{re}}(\mathbf{v}) = \mathbf{M} \cdot \mathbf{v} \tag{5.13}$$

at the microphone locations. As the analytic calculation of the pressure results in a complex vector $\mathbf{p}_{\text{re}}(\mathbf{v})$, many solutions are possible whose absolute values match the given target directivity. Therefore, two strategies of matching the pressure values are evaluated: assuming a constant phase of zero for all microphones and using nonlinear optimization to find a solution with arbitrary phase whose absolute values match the target.

The first method regards a linear problem that is solved efficiently by matrix inversion using a multiplication with the pseudoinverse of matrix $\mathbf{M}$, minimizing the residual (cf. Eq. (2.78))

$$\mathbf{r}_{\text{p}} = \mathbf{p}_{\text{re}}(\mathbf{v}) - \mathbf{p}_{\text{mic}}. \tag{5.14}$$

The directivity pattern of the spherical loudspeaker is thus not matched in the spherical harmonics domain but in the spatial domain at the discrete sampling points of the spherical microphone array.

The second approach offers a higher degree of freedom, with the possibility to use any phase as long as the magnitudes values of the pressure are equal. The modified residual can be defined as

$$\mathbf{r}_{\text{abs}} = |\mathbf{p}_{\text{re}}(\mathbf{v})| - \mathbf{p}_{\text{mic}} \tag{5.15}$$

describing the deviation from a perfect magnitude match of the two functions at the microphones. As the absolute value in the calculation of the residual regards a nonlinear computation, the computation with a generalized matrix inversion is not feasible. A least-squares optimization approach is followed to obtain solutions for the optimal excitation. The phase at the position of the microphones is thus considered arbitrary and optimized for a best possible match.

The results of the directivity synthesis are depicted in Fig. 5.10 and Fig. 5.11 for the radiation of a trumpet and a violin, respectively. Both methods described have been evaluated, assuming a constant phase in the left column and optimizing the phase in the right column. The shade of gray of the round spots represents

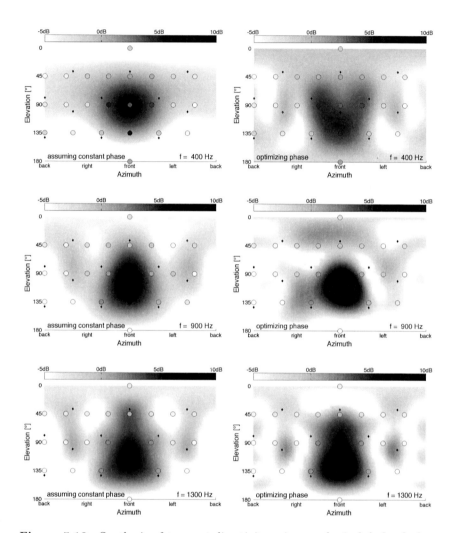

Figure 5.10.: Synthesis of trumpet directivity using a spherical dodecahedron
loudspeaker with individual transducer gain control. The round
spots represent the directivity value of the musical instrument,
the black dots show the location of the loudspeaker membranes.
Simulation performed using a constant phase (left) and an optimized phase (right) at frequencies of 400 Hz, 900 Hz and 1300 Hz.
[Pol09c]

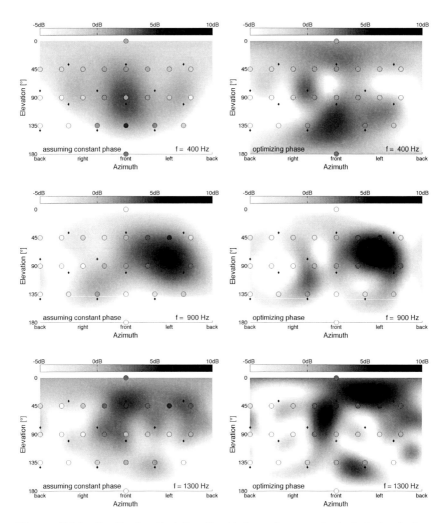

Figure 5.11.: Synthesis of violin directivity using a spherical dodecahedron loudspeaker with individual transducer gain control. The round spots represent the directivity value of the musical instrument, the black dots show the location of the loudspeaker membranes. Simulation performed using a constant phase (left) and an optimized phase (right) at frequencies of 400 Hz, 900 Hz and 1300 Hz. [Pol09c]

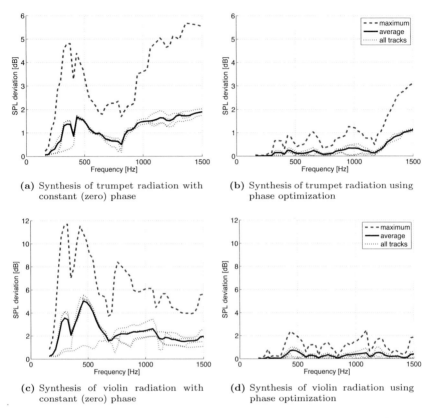

(a) Synthesis of trumpet radiation with constant (zero) phase

(b) Synthesis of trumpet radiation using phase optimization

(c) Synthesis of violin radiation with constant (zero) phase

(d) Synthesis of violin radiation using phase optimization

Figure 5.12.: Average and maximum SPL deviation between measured instrument directivity and simulated dodecahedron directivity [Pol09c]

the logarithmic directivity of the musical instrument in that direction, while the black dots represent the center points of the membranes of the dodecahedron loudspeaker. The continuos grayscale in the background is the resulting SPL caused by the simulated radiation of the loudspeaker. Identical contrast of the spots and the background shading thus represents a perfect match at that microphone position. The frequencies 400 Hz, 800 Hz and 1300 Hz are depicted in the figures.

It can be seen, that the match is generally higher for higher degrees of freedom using the arbitrarily optimized phase. Even maxima in-between the membrane center points can be matched well in that case (cf. Fig. 5.10 at 1300 Hz) while forcing a fixed phase of zero from the averaged data yields higher deviations, especially for highly focused sources.

The error plots in Fig. 5.12 confirm these observations. For each instrument the frequency dependent average SPL error and the maximum SPL error considering all microphones are depicted. The dotted lines show the average error for a similar target directivity obtained using different track recordings of the musical instruments. In the operating range of the mid-range unit of the spherical dodecahedron loudspeaker, maximum deviations from mostly under 2 dB occur for the optimized phase, while for the zero-phase-approach deviations up to 6 dB and 12 dB occur for the synthesis of the trumpet and the violin, respectively.

The feeding voltages to obtain the required surface velocities for the membranes can be obtained either by a laser Doppler vibrometer or by the electroacoustic equivalent circuit diagram of the mechanical elements involved [Pol09c].

5.4.2. Directivity synthesis for RIR measurements: Sequential approach using tilted dodecahedron loudspeaker

In this section the directivity pattern of a technical target sound source is synthesized using the measured radiation patterns of a spherical loudspeaker array. The synthesis result is a weighting vector that can also be applied to room impulse responses (RIRs) measured with the spherical loudspeaker. Assuming linear and time-invariant (LTI) systems this approach allows to synthesize the room impulse responses for arbitrary target source directivity up to a specific spatial resolution.

As experimental evaluation the tilted dodecahedron loudspeaker as depicted in Fig. 5.4b is used. The device is mounted on a turntable to increase the spatial resolution to $L = 120$ (virtual) transducers using ten azimuthal orientations in steps of $\Delta\phi = 36°$ [Kun11]. The radiation of all membranes of the tilted dodecahedron loudspeaker in all positions measured at distance r can be described by the matrix

$$\hat{\mathbf{P}} = [\hat{\mathbf{p}}_1, \hat{\mathbf{p}}_2, \ldots, \hat{\mathbf{p}}_L] \, . \tag{5.16}$$

containing the spherical wave coefficients (SWS) of the membranes' directivity. The transfer functions of unit excitation (re 1 V) of the lth transducer are given as vector $\hat{\mathbf{p}}_l$ in the SH domain. Applying the spherical harmonic transform (SHT, cf. Eq. (2.75)) a set of measured sound pressure values obtained on a spherical sampling scheme can be used to calculate the vector

$$\hat{\mathbf{p}}_l = \text{vec}_{\text{SH}}\{p_{nm,l}\} \tag{5.17}$$

describing the SWS of the sound pressure measured at the distance r around the focal point (center of the spherical loudspeaker). For the measurement the Gaussian quadrature sampling scheme (cf. Sec. 2.5.2) has been used as it offers a exact SHT for order limited functions without the need of explicit matrix inversion.

As target the directivity of a cubic loudspeaker (as depicted in Fig. 5.14a with an edge length of $12\,\text{cm}$) that is represented by its SWS as vector $\hat{\mathbf{p}}_t$ has been used. The input voltage vector for the spherical sound source with L transducers contains the complex input voltages u_l for $l = 1 \ldots L$ denoted as

$$\mathbf{u} = [u_1, u_2, \ldots, u_L]^T . \tag{5.18}$$

The resulting directivity pattern can be computed using the equation

$$\hat{\mathbf{p}} = \hat{\mathbf{P}}\mathbf{u} \tag{5.19}$$

with $\hat{\mathbf{p}}$ being the sound pressure of the synthesis result. This superposition approach holds for coherent excitation of only one driver at a time, avoiding crosstalk effects that impact the result. Depending on the application either measured or simulated directivity patterns can be used as target patterns. In this example the measurement result of the directivity pattern of the cubical loudspeaker is used as target directivity, as it can be measured to verify the obtained results.[9]

A solution for the input voltage vector $\mathbf{u}$ can be formulated by least-mean squares minimization of the deviation of target and synthesized directivity in the spherical harmonic domain [Pol11b]:

$$\mathbf{u}_t = \underset{\mathbf{u} \in \mathbb{C}^L}{\arg\min} \left\| \hat{\mathbf{p}}_t - \hat{\mathbf{P}}\mathbf{u} \right\|_2^2 \tag{5.20}$$

---

[9]Here the assumption is made that the directivity is uniquely defined for every frequency. In reality, however, the directivity of natural sound sources also depends on different properties as style and strength of excitation as illustrated in Chapter 3.

Numerical solutions of this equation can be found by using the Moore-Penrose pseudoinverse. In this example Tikhonov regularization has been applied giving preference to solutions with smaller norm as

$$\mathbf{u}_t = \left(\hat{\mathbf{P}}^H \hat{\mathbf{P}} + \varepsilon \mathbf{I}\right)^{-1} \hat{\mathbf{P}}^H \hat{\mathbf{p}}_t \qquad (5.21)$$

with $\varepsilon = 10^{-3}$ being the regularization parameter in the given example and $\mathbf{I}$ being the $L \times L$ Identity matrix [Pol11b]. To determine the RIR for the derived excitation vector $\mathbf{u}_t$ the RIRs of all L membranes are measured individually and are stored in the row vector

$$\mathbf{h} = [h_1, h_2, \ldots, h_L] \qquad (5.22)$$

with

$$h_l = p_l(\mathbf{u}) \quad \text{for} \quad u_{l'} = \delta_{ll'} \qquad (5.23)$$

being the RIR as measured for transducer position l. The Kronecker delta $\delta_{ll'}$ has been defined in Eq. (2.33). The measurement time of a RIR for one orientation of the spherical array and the reference loudspeaker requires approx. 30 seconds. Hence, ten sphere orientations corresponding to $L = 120$ membrane orientations require roughly five minutes of measurement time.

Applying superposition of the elements of $\mathbf{h}$ with the calculated input voltage vector $\mathbf{u}_t$ derived from Eq. (5.21) yields the room impulse response

$$h_t = \mathbf{h}\mathbf{u}_t \qquad (5.24)$$

for the excitation of a target directivity pattern $\hat{\mathbf{p}}_t$.

In order to evaluate the proposed method a comparative measurement has been conducted in a lecturing hall with an approx. size of $8\,\text{m} \times 6\,\text{m} \times 3\,\text{m}$, a mean reverberation time of approximately $T_{60} = 0.9\,\text{s}$ and a distance of source and receiver of approx. $4.8\,\text{m}$. The superposition weights are calculated for directivity synthesis and then applied to measured RIRs of single membrane excitation. A band-pass filter from $400\,\text{Hz}$ to $4\,\text{kHz}$ is applied to the measurement results as this has shown to be the operating range for this measurement setup [Kun11]. Between $300\,\text{Hz}$ and $1.5\,\text{kHz}$ very high correlation values between original RIR and the synthesis result can be derived [Kun12]. In Fig. 5.13 the room transfer function (RTF) is plotted as measured with the target directivity and synthesized from the measurements of the tilted dodecahedron loudspeaker. In Fig. 5.14 the room impulse responses are plotted on a logarithmic scale in time domain.

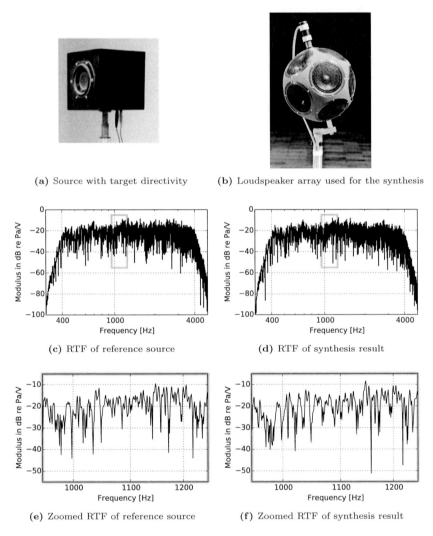

(a) Source with target directivity

(b) Loudspeaker array used for the synthesis

(c) RTF of reference source

(d) RTF of synthesis result

(e) Zoomed RTF of reference source

(f) Zoomed RTF of synthesis result

Figure 5.13.: Measured RTF using a reference source and the synthesized result using the loudspeaker array. The zoomed part of the RTF is marked by a frame.

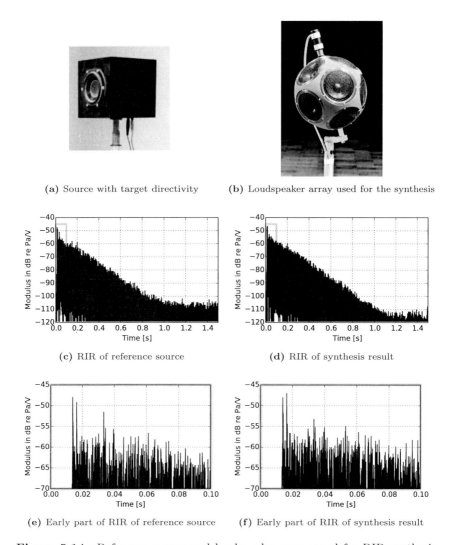

(a) Source with target directivity   (b) Loudspeaker array used for the synthesis

(c) RIR of reference source   (d) RIR of synthesis result

(e) Early part of RIR of reference source   (f) Early part of RIR of synthesis result

Figure 5.14.: Reference source and loudspeaker array used for RIR synthesis and results in large and small scale. The zoomed part of the RIR is marked by a frame.

5.4.3. Directivity synthesis for high resolution radiation patterns: Sequential approach using an optimized spherical loudspeaker array

In this section the results of the directivity synthesis using measured pressure functions of the optimized spherical sound source as described in Sec. 5.3.2 are visualized. The 28 transducers of the array system have been simulated in 24 positions resulting in 672 membrane orientations. The individual behavior of the transducer is taken into account, as the measured directivity pattern of the individual transducers are used for the computation.

Giving arbitrary target directivity patterns $\hat{\mathbf{p}}_t$, solutions of the ideal excitation vector $\mathbf{u}_t$ can be found by multiplication with the pseudoinverse matrix or by using inversion with order dependent Tikhonov regularization as

$$\mathbf{u}_t = \hat{\mathbf{P}}^H \left(\hat{\mathbf{P}} \hat{\mathbf{P}}^H + \varepsilon \mathbf{D} \right)^{-1} \hat{\mathbf{p}}_t. \tag{5.25}$$

The diagonal matrix $\mathbf{D}$ defined as in Eq. (2.85) has been used. In the example given here the regularization constant is set to $\varepsilon = 10^{-6}$, the maximum order chosen as $n_{\max} = \lceil 1.5 \cdot kr \rceil$ or $n_{\max} = 70$ (whatever value is smaller). These parameters yield accurate results for the directivity synthesis.

As an example three target directivity patterns have been synthesized: the ideal Dirac delta function on the sphere as depicted on page 111 and a set of simulated and measured HRTFs of artificial heads (cf. Sec. 4.1) depicted on page 112 and 113, respectively. Each pattern has been synthesized using both matrix inversion by Moore-Penrose pseudoinverse and using order dependent Tikhonov regularization, cf. Sec. 2.4.4.

It can be observed that with regularization smoother patterns are obtained that use significantly smaller amplitudes in the superposition vector. The (unrealistic) narrow beam at low frequencies is approximated at the cost of noise in all directions for the pseudoinverse, and at the cost of a wider lobe for the regularization approach.[10] The horizontal ripple that is a visible artifact in the measured HRTFs at 500 Hz in Fig. 5.21 is reconstructed using the pseudoinverse. Using regularization the synthesis result of the 672 individual transducer directivity patterns in Fig. 5.23 seems like the perfect correction of the original measurement result without this horizontal ripple. For frequencies starting from 1000 Hz

---

[10]Note that the amplitudes of the beams vary with the frequency, as the mean square error of the deviations between the ideal beam and the generated beam are minimized for all directions.

the deviations between original and synthesized directivity pattern become visible for the order dependent Tikhonov regularization.

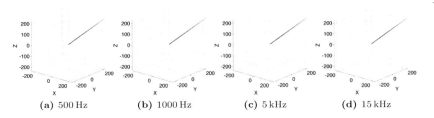

(a) 500 Hz   (b) 1000 Hz   (c) 5 kHz   (d) 15 kHz

Figure 5.15.: Target function: Ideal narrow beam

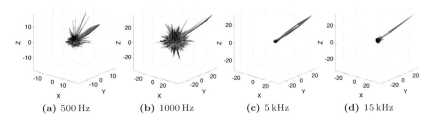

(a) 500 Hz   (b) 1000 Hz   (c) 5 kHz   (d) 15 kHz

Figure 5.16.: Synthesis of the narrow beam from measured loudspeaker directivity patterns using inversion by Moore-Penrose pseudoinverse

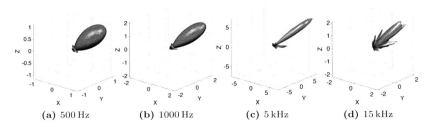

(a) 500 Hz   (b) 1000 Hz   (c) 5 kHz   (d) 15 kHz

Figure 5.17.: Synthesis of the narrow beam from measured loudspeaker directivity patterns using inversion with order dependent Tikhonov regularization

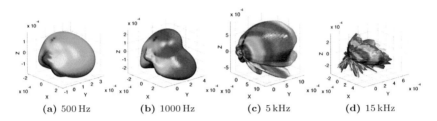

(a) 500 Hz   (b) 1000 Hz   (c) 5 kHz   (d) 15 kHz

Figure 5.18.: Target function: BEM simulation of HRTFs

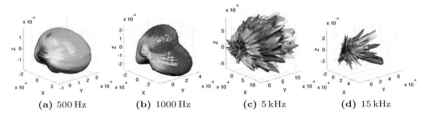

(a) 500 Hz   (b) 1000 Hz   (c) 5 kHz   (d) 15 kHz

Figure 5.19.: Synthesis of the simulated HRTFs from measured loudspeaker directivity patterns using inversion by Moore-Penrose pseudoinverse

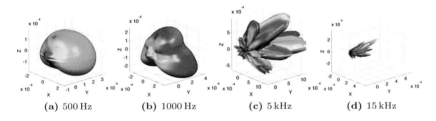

(a) 500 Hz   (b) 1000 Hz   (c) 5 kHz   (d) 15 kHz

Figure 5.20.: Synthesis of the simulated HRTFs from measured loudspeaker directivity patterns using inversion with order dependent Tikhonov regularization

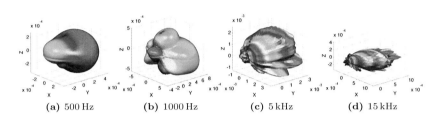

(a) 500 Hz   (b) 1000 Hz   (c) 5 kHz   (d) 15 kHz

Figure 5.21.: Target function: Measured HRTFs

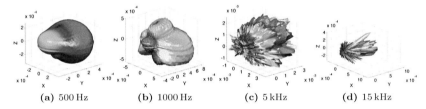

(a) 500 Hz   (b) 1000 Hz   (c) 5 kHz   (d) 15 kHz

Figure 5.22.: Synthesis of the measured HRTFs from measured loudspeaker directivity patterns using inversion by Moore-Penrose pseudoinverse

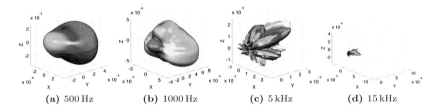

(a) 500 Hz   (b) 1000 Hz   (c) 5 kHz   (d) 15 kHz

Figure 5.23.: Synthesis of the measured HRTFs from measured loudspeaker directivity patterns using inversion with order dependent Tikhonov regularization

5.5. Summary directivity patterns in room acoustic measurements

In this chapter the tools have been derived to obtain room acoustical measurements with respect to arbitrary source directivity. An analytic model for spherical sound sources is given, allowing to calculate the resulting sound pressure in the entire room from a given set of membrane velocities on a spherical loudspeaker chassis. This model has been used to analyze the performance of existing loudspeaker arrays and to design specialized loudspeaker arrays for the purpose of efficient directivity synthesis. The mathematical framework presented earlier allows rapid analytical calculation of the radiation of the array, which can be optimized for simultaneous synthesis or sequential synthesis methods (using a subsequent azimuthal rotation of the array). Using these devices for directivity synthesis the measured directivity pattens for all individual membranes can be used, taking into account the deviations between analytical model and physical reality. Following this approach the individual behavior such as the specific sensitivities of the transducers are considered, thus increasing the accuracy of synthesis.

Finally, three examples for the use of these devices in room acoustical measurements have been given:

The first example shows the performance of directivity synthesis using a multichannel dodecahedron loudspeaker to match the target data as obtained on a spatially coarse sampling grid. As the target pattern is given as amplitudes, a nonlinear optimization of the phases allows to improve the performance of the directivity synthesis.

The second example shows the RIR synthesis for a target directivity pattern using a spherical sound source and a sequential synthesis approach, showing a good match in the range of the predicted operating frequencies.

The last example visualizes the performance of the different matrix inversion approaches on the resulting synthesized directivity patterns. Depending on the used inversion a focus can be set to derive high similarity, lower amplitudes of the synthesis vector and/or a smoothed target directivity.

6

Including directivity patterns in room acoustic simulations

In this chapter methods and algorithms are described that allow the integration of arbitrary source and receiver directivity patterns in room acoustic simulation software. The simulation methods for room acoustics can be separated in wave based methods and particle based methods. While wave based simulations yield precise solutions for known geometries and known boundary conditions, the computational cost are high, especially with rising frequency.

Modern simulation programmes use hybrid simulation methods [Sch11], combining e.g. the precise but computationally expensive image source method [All79] with a ray tracing approach for the late reflections. Both of these particle based methods can be extended to include directivity patterns in a straightforward manner, multiplying the outgoing or incoming sound with the specific directivity value in the direction of sound radiation or reception. Measured or simulated directivity patterns can be taken into account in these algorithms (either as a complex function or using magnitude values) in space domain or the spherical Fourier domain. The latter offers implicit interpolation of the directivity patterns, thus allowing to retrieve the value for an arbitrary direction without the need for explicit interpolation from neighboring points.

For wave based simulations the effect of directivity patterns can be included if the modal structure of the considered room is known. Arbitrary patterns can be represented as a specific set of physical multipoles whose impact on the room is calculated by taking the spatial derivatives in Cartesian coordinates of the eigenfunctions of the room. The obtained description in terms of physical multipoles can then be converted to represent directivity patterns of arbitrary SWS, both for sender and/or receiver side.

Parts of this chapter have been published in [Pol13b] and [Pel12a].

6.1. Directivity in wave based room simulation

Different techniques can be used to obtain reliable results from room acoustical simulations. The wave based numerical approaches are regarded as precise
methods with high computational cost (especially for higher frequencies) and
can be performed using e.g. the Boundary Element Method (BEM), the Finite
Element Method (FEM) or the Fast Multipole Method (FMM). In practice, the
computational demand of these simulation methods is often too high for a realistic room simulation of complex scenarios over the entire audible frequency
range. For simpler geometries the room modes and thus the room transfer function can be derived analytically. The room transfer function for these simple
geometries can be computed from the summation of the calculated room modes
[Kut00; Pie89]. The resulting *room transfer function* (RTF) is valid for both
perfectly omnidirectional sources and receivers. The Cartesian derivatives of the
eigenfunctions of a particular room allows to exchange the solution for monopole
sources and receivers to a solution for dipole sources and/or receivers, whereas
multiple derivatives yield the result for the physical multipoles of higher orders.
This multipole description can then be converted to a representation in terms
of the spherical wave spectrum (SWS), allowing to determine the analytically
computed room impulse response for arbitrary source and receiver directivities
in rooms of known and differentiable eigenmodes.[1]

In order to validate the method a simple rectangular room (shoebox shaped
geometry) is considered and the transfer path between two arbitrary points in
the room is computed with respect to the arbitrary given directivity patterns of
source and receiver. The result is compared with a numerical BEM simulation of
the same geometry, using *LMS Virtual.Lab* witch offers the feature to compute
the room response for dipole and quadrupole excitation.

6.1.1. Spatial sampling of the eigenfunctions of a room

The eigenfrequencies of a rectangular room with rigid walls can be written as

$$\omega_i = c\pi\sqrt{\left(\frac{n_{x,i}}{L_x}\right)^2 + \left(\frac{n_{y,i}}{L_y}\right)^2 + \left(\frac{n_{z,i}}{L_z}\right)^2} \tag{6.1}$$

---

[1]Note that in order to be in compliance with the conventions used by Gumerov et al.
[Gum03] a different definition for the spherical harmonics are used in this section as follows:
$Y_n^m(\theta,\phi) = (-1)^m\sqrt{\frac{2n+1}{4\pi}\frac{(n-|m|)!}{(n+|m|)!}}P_n^{|m|}(\cos\theta)e^{jm\phi}$.

with the modal numbers $n_{x,i}$, $n_{y,i}$ and $n_{z,i}$ being all possible combinations of non-negative integer values and c being the speed of sound [Kut00; Mec08].

The formula for the modal superposition in a room reads as

$$H(\omega) = -\frac{4\pi c^2}{V} \sum_i \frac{\psi_i(\mathbf{r_s})\psi_i(\mathbf{r_r})}{(\omega^2 - \omega_i^2 - \mathrm{j}\delta_i\omega_i)K_i} \qquad (6.2)$$

with V being the volume of the rectangular room of dimensions $L_x \times L_y \times L_z$, δ_i being the damping constant (here $\delta_i = 0$ is assumed, representing an undamped room) and $K_i = \iiint \psi_i^2(\mathbf{r})\,\mathrm{d}V$ being a normalization constant for the eigenmodes [Kut00]. The result constitutes the frequency response of the room transfer path from a point source at $\mathbf{r_s}$ to a receiver point at $\mathbf{r_r}$. Applying the inverse Fourier transform this term directly states the room impulse response for the given transfer path.

The eigenfunctions in the room sampled at a discrete position $\mathbf{r} = (x, y, z)$ can be expressed as frequency response

$$\psi_i(\mathbf{r}) = \cos\left(\pi n_{x,i}\frac{x}{L_x}\right)\cos\left(\pi n_{y,i}\frac{y}{L_y}\right)\cos\left(\pi n_{z,i}\frac{z}{L_z}\right). \qquad (6.3)$$

For any arbitrary room an infinite number of modes exist, while in practice an upper frequency limit is defined to restrict the number of modes to calculate. Inserting these spatially sampled eigenfunctions for both source and receiver positions in Eq. (6.2), yields the room transfer function for point sources and point receivers with omnidirectional radiation and omnidirectional sensitivity, respectively.

6.1.2. Radiation of monopoles and physical multipoles

The free-field sound radiation of a monopole source located in the origin of the coordinate system can be formulated as

$$p(\mathbf{r}) = \mathrm{j}\rho_0 ck Q_s G(\mathbf{r}|\mathbf{0}) \qquad (6.4)$$

with $G(\mathbf{r}|\mathbf{r}')$ being the Green's function, $\mathbf{r}'$ the source position and $\mathbf{r}$ the observation point [Wil99]. The radiation of a dipole (here exemplarily oriented in x direction) can be described as

$$p(\mathbf{r}) = \mathrm{j}\rho_0 ck Q_s d\frac{\partial}{\partial x}G(\mathbf{r}|\mathbf{0}). \qquad (6.5)$$

This term differs from the radiation of a monopole by a multiplication of the dipole distance d and by replacing the Green's function by its Cartesian derivative [Wil99]. The physical multipoles of higher orders can thus composed by multiple differentiations in Cartesian coordinates.

6.1.3. Spherical wave spectrum of physical multipoles

To convert the solution in terms of physical multipoles to a description as spherical wave spectrum (SWS) all possible combinations of Cartesian derivatives of the eigenfunctions are calculated and sampled at the position of source and receiver. The spherical wave coefficients for all physical multipoles up to a specific order are calculated and stored as SWS in a matrix. This matrix describes an underdetermined system of equations, so multiple solutions exist to represent a directivity pattern given as SWS by a set of physical multipoles. The transformation matrix is sparsely populated and a solution of the inverse problem can be found, e.g. by LU decomposition. This calculation is not unique, so it is possible to speed up the computation for spherical wave spectra of higher orders, as only a subset of the derivatives have to be computed to find a set of multipoles for any arbitrary directivity pattern.

Any compact sound radiator can be described by the multipole expansion c_{nm} as formulated in Sec. 2.3.4 as superposition of the fundamental singular solutions of the Helmholtz equation. Defining $c_{nm}^{(n_x,n_y,n_z)}$ as the multipole expansion of the monopole point source after multiple Cartesian differentiation and using the abbreviation

$$S_n^m(\mathbf{r}) = h_n(kr)Y_n^m(\theta, \phi) \tag{6.6}$$

with the vectorized position $\mathbf{r} = \vec{e}_r\, r + \vec{e}_\theta\, \theta + \vec{e}_\phi\, \phi$ yields:

$$\left(\frac{\partial}{\partial x}\right)^{n_x}\left(\frac{\partial}{\partial y}\right)^{n_y}\left(\frac{\partial}{\partial z}\right)^{n_z} p(\mathbf{r}) = \tag{6.7}$$

$$= \left(\frac{\partial}{\partial x}\right)^{n_x}\left(\frac{\partial}{\partial y}\right)^{n_y}\left(\frac{\partial}{\partial z}\right)^{n_z}\sum_{n=0}^{\infty}\sum_{m=-n}^{n} c_{nm}^{(0,0,0)} S_n^m(\mathbf{r}) \tag{6.8}$$

$$= \sum_{n=0}^{\infty}\sum_{m=-n}^{n} c_{nm}^{(n_x,n_y,n_z)} S_n^m(\mathbf{r}) \tag{6.9}$$

For a point source the multipole expansion can be given as

$$c_{nm}^{(0,0,0)} = \delta_{n0}\delta_{m0}\frac{\rho_0 c k^2}{\sqrt{4\pi}} Q_s. \tag{6.10}$$

Gumerov et al. [Gum03] give the Cartesian derivatives of the singular solutions of the Helmholtz equation as:

$$\frac{\partial}{\partial x} S_n^m(\mathbf{r}) = \frac{k}{2} \left[T_{n,1}^m(\mathbf{r}) + T_{n,2}^m(\mathbf{r}) \right] \tag{6.11}$$

$$\frac{\partial}{\partial y} S_n^m(\mathbf{r}) = \frac{\mathrm{j}k}{2} \left[T_{n,2}^m(\mathbf{r}) - T_{n,1}^m(\mathbf{r}) \right] \tag{6.12}$$

$$\frac{\partial}{\partial z} S_n^m(\mathbf{r}) = k\, T_{n,3}^m(\mathbf{r}) \tag{6.13}$$

The abbreviations $T_{n,1}^m(\mathbf{r})$, $T_{n,2}^m(\mathbf{r})$ and $T_{n,3}^m(\mathbf{r})$ are defined as

$$T_{n,1}^m(\mathbf{r}) = b_{n+1}^{-m-1} S_{n+1}^{m+1}(\mathbf{r}) - b_n^m S_{n-1}^{m+1}(\mathbf{r}) \tag{6.14}$$

$$T_{n,2}^m(\mathbf{r}) = b_{n+1}^{m-1} S_{n+1}^{m-1}(\mathbf{r}) - b_n^{-m} S_{n-1}^{m-1}(\mathbf{r}) \tag{6.15}$$

$$T_{n,3}^m(\mathbf{r}) = a_{n+1}^m S_{n-1}^m(\mathbf{r}) - a_n^m S_{n+1}^m(\mathbf{r}) \tag{6.16}$$

with

$$a_n^m = \begin{cases} \sqrt{\dfrac{(n+1+|m|)(n+1-|m|)}{(2n+1)(2n+3)}} & \text{for} \quad n \geq |m| \\ 0 & \text{for} \quad n < |m| \end{cases} \tag{6.17}$$

and

$$b_n^m = \begin{cases} \sqrt{\dfrac{(n-m-1)(n-m)}{(2n-1)(2n+1)}} & \text{for} \quad 0 \leq m \leq n \\ -\sqrt{\dfrac{(n-m-1)(n-m)}{(2n-1)(2n+1)}} & \text{for} \quad -n \leq m < 0 \, . \\ 0 & \text{for} \quad |m| > n \end{cases} \tag{6.18}$$

The sound pressure level for a multipole source that is equivalent to the ith, kth and lth derivation in x, y and z direction, respectively, is:

$$p^{(i,k,l)}(\mathbf{r}) = \left[\frac{\partial}{\partial x} \right]^i \left[\frac{\partial}{\partial y} \right]^k \left[\frac{\partial}{\partial z} \right]^l p^{(0,0,0)}(\mathbf{r}) \tag{6.19}$$

$$= \left[\frac{\partial}{\partial x} \right]^i \left[\frac{\partial}{\partial y} \right]^k \left[\frac{\partial}{\partial z} \right]^l c_{00}^{(0,0,0)} S_0^0(\mathbf{r}) \tag{6.20}$$

$$= \sum_{n=0}^{\infty} \sum_{m=-n}^{n} c_{nm}^{(i,k,l)} S_n^m(\mathbf{r}) \tag{6.21}$$

A linearization of the SWS as described in Sec. 2.4.1 allows to use matrix notation for the conversion of the physical multipoles into the multipole expansion given as SH coefficients as

$$\mathbf{c}^{(i,k,l)} = \text{vec}_{\text{SH}} \left\{ c_{nm}^{(i,k,l)} \right\}. \tag{6.22}$$

Combining all multiple partial derivatives of the radiated sound pressure of a point source with a maximum number of up to n_{max} derivatives, this yields the transformation matrix

$$\mathbf{M} = \left\{ \mathbf{c}^{(i,k,l)} \quad \forall \quad i+k+l \leq n_{\text{max}} \right\} \tag{6.23}$$

$$= \left[\mathbf{c}^{(0,0,0)} \ \mathbf{c}^{(1,0,0)} \ \mathbf{c}^{(0,1,0)} \ \dots \ \mathbf{c}^{(0,0,n_{\text{max}})} \right] \tag{6.24}$$

of size $(n_{\text{max}}+1)^2 \times \sum_{n=0}^{n_{\text{max}}} \frac{(2+n)!}{2 \cdot n!}$ constituting an underdetermined system of equations for SH orders greater than one.

Any source directivity given by the coefficients c_{nm} of its SWS can be represented by superposing the partial derivatives of the monopole source as

$$\mathbf{c} = \mathbf{Md} \tag{6.25}$$

with $\mathbf{d}$ being the coefficient vector for all computed partial derivations, which can be calculated e.g. by an LU decomposition of the matrix $\mathbf{M}$.

With the help of the room impulse responses for the partial derivatives as formulated in Eq. (6.11) to Eq. (6.13) it is thus possible to superpose the room impulse response for any sound source with arbitrarily given SWS.

6.1.4. Verification of the algorithm

In order to verify the proposed analytic model a numerical simulation is carried out using the *Boundary Element Method* (BEM) of *LMS Virtual.Lab* with a suitable mesh to simulate a shoebox type room up to a frequency of 1000 Hz. The room is chosen to have the dimensions 80 cm × 50 cm × 30 cm as depicted in Fig. 6.1, with the source placed at $\mathbf{r_s} = (15\,\text{cm}, 15\,\text{cm}, 15\,\text{cm})$ and the receiver placed in one of the room corners at $\mathbf{r_r} = (0,0,0)$. While the receiver is kept with omnidirectional sensitivity, the directivity of the source is varied by modification of the eigenfunctions of the room. The analytic calculation is performed up to a frequency of 10 kHz.

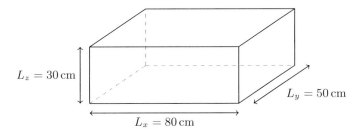

Figure 6.1.: Rectangular room with solid walls used for the calculation of the modes and its RTF with respect to varying directivity patterns

The result for the analytic calculation of the monopole source is compared with the BEM results in Fig. 6.2. It can be seen that both simulation results match precisely in the evaluated frequency range up to 1 kHz (which regards the upper limit for the BEM computations due to computational constraints). All peaks and notches of the room transfer function equal both in frequency and absolute levels.

Replacing the point source in the analytical model with a dipole source by using the first Cartesian derivative of the eigenfunctions as

$$\psi_{i,\mathrm{dipole}}(\mathbf{r_s}) = \frac{\partial}{\partial x}\psi_i(\mathbf{r_s}) \tag{6.26}$$

allows to calculate the frequency response of a dipole source to a monopole receiver. As Eq. (6.1) is always valid, the eigenfrequencies are identical to the previous case, while some of the modes are not excited due to the dipole character of the source, i.e. all modes with the mode number $n_{x,i} = 0$ are not being excited when computing the response for the dipole source by applying the first derivative in x-direction. A comparison with the BEM simulation of that dipole source as depicted in Fig. 6.3 shows a good match with minor deviations at very low frequencies below the first room resonance of less than 1 dB. The used BEM simulation software allows the use of explicit dipole sources, which provides the same result as the BEM simulation of two monopoles of opposite phase in close vicinity multiplied with their distance.

Applying the second derivative to the eigenfunction of the monopole yields the longitudinal quadrupole for the source directivity pattern as

$$\psi_{i,\mathrm{quadrupole}}(\mathbf{r_s}) = \frac{\partial^2}{\partial x^2}\psi_i(\mathbf{r_s}) \tag{6.27}$$

121

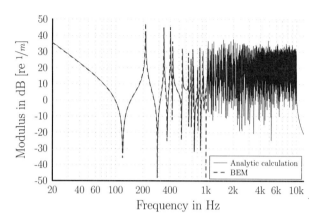

Figure 6.2.: Monopole source and receiver: Comparison of frequency response functions obtained by analytic model and by numeric simulation [Pol13b].

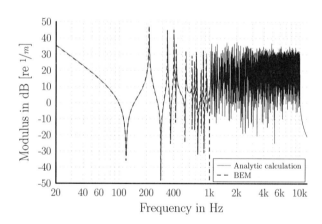

Figure 6.3.: Dipole source in x-direction and monopole receiver: Comparison of frequency response functions obtained by analytic model and by numeric simulation [Pol13b].

Inserted in Eq. (6.2) and compared with the corresponding BEM simulation result shows a match in both frequencies and absolute levels at the resonances, but with significant differences in-between these peaks. This resulting transfer function varies for small variations of the upper frequency limit of the analytic simulation, suggesting a significant influence of the higher modes even in the low

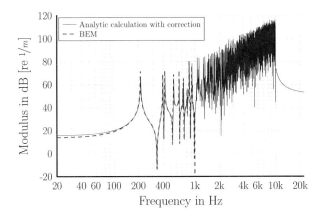

Figure 6.4.: Longitudinal quadrupole source in x-direction and monopole receiver: Comparison of frequency response functions obtained by analytic model with offset correction and by numeric simulation [Pol13b].

frequency range far from their actual resonance peaks due to hard truncation at a specific frequency. Correcting the result of the analytic calculation by the subtraction of a frequency independent offset, yields a good match with the BEM simulation results as depicted in Fig. 6.4. This compensation is performed by calculating the frequency response for the first few modes starting at low frequencies and matching the residual at DC for the full simulation up to higher modes as described by Dietrich [Die13].

6.2. Directivity in particle based room simulation

In order to obtain faster results over a broad frequency spectrum and complex room geometries often ray based simulations are used. Current state-of-the-art software uses a hybrid method, combining e.g. the (deterministic) image source method [All79] for the direct sound and early reflections with an efficient statistical path based approach such as *ray tracing* or related techniques such as *beam-tracing, cone-tracing* and *radiosity* [Kut00; Sch11; Kou12].

Both deterministic and stochastic methods allow to include the directivity of source and receiver in a straightforward manner. The directional dependence of the sensitivity can be implemented in ray tracing by a scaling factor for each emitted ray at the sender side and by a directional sensitive detection sphere on the receiver side. As the approach is energy based the magnitude values of the directivity patterns are sufficient.

Also the image source model can be extended using the directional sensitivity of source and receiver. Here complex directivity patterns (including phase information) can be implemented, which are mirrored together with the source position in the computation. Working with the spherical wave spectrum of the directivity functions, the interpolation of the spherical functions to any arbitrary directions can be performed efficiently.

6.2.1. Implementation for fixed source and receiver directivity patterns

In common geometrical acoustics models, sources can be modeled as point sources with optional directivity. For most of the natural sound sources this is a valid simplification as long as the source can be considered as small from the listening point. Most symphonic instruments fulfill this requirement when perceived from the audience [Pel12a].

The directivity pattern of a receiver is implemented using energy histograms in a general frequency-, time- and direction dependent data structure. A directional dependent sensitivity can be applied to the simulation, so that the acoustical transfer paths are simulated for the correct directivity patterns of source and receiver. The directivity patterns are currently being used in third octave bands in ITA's in-house development RAVEN [Sch11].

6.2.2. Implementation for exchangeable directivity patterns

For an alternative implementation a set of directivity functions is calculated which allows the superposition of any receiver directivity pattern up to a maximum spherical harmonic (SH) order. The spherical harmonics as defined in Eq. (2.29) regard a set of orthonormal basis functions and are thus perfectly suited to span the function space of arbitrary directivity patterns. As the simulations are performed in time-domain, the use of a real-valued set of spherical harmonic functions as used in Zotter [Zot09a] is more efficient. The obtained simulation results can be used for arbitrary directivity patterns that can be changed efficiently in the post-processing stage.

For directivity patterns represented as spherical wave spectra as defined in Sec. 2.3.3, the property of computationally cheap azimuthal variation can be exploited: Implementing an azimuthal rotation of static directivity patterns (a common use case for musical instruments and human listeners) the Wigner-D matrix as defined in Eq. (2.49) simplifies to

$$D_{k,m}^{(n)}(\alpha, 0, 0) = \mathrm{e}^{-\mathrm{j}k\alpha} \qquad (6.28)$$

and can be used to calculate rotations around the z-axis orders of magnitude faster than arbitrary rotations around a center point.

For the implementation of both source and receiver directivity pattern the room acoustic transfer path is depicted in Fig. 6.5. The vectorized SWS of the source and receiver can both be individually rotated by Wigner-D rotation matrices. The room can then be described as a transformation matrix mapping the SWS of the source to the SWS of the receiver at the fixed positions the simulation has been computed for.

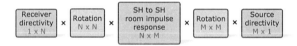

Figure 6.5.: Signal chain to simulate a transfer path for arbitrary source and receiver directivity patterns in RAVEN [Pel12b]

6.2.3. Efficient pre-processing for exchangeable receiver directivity patterns

For an immersive real-time auralization all computationally expensive operations can be done in a pre-processing step. This concerns mainly the simulation of spatial room impulse responses as well as the convolution of several minutes of recorded solo instruments with long room impulse responses for all combinations of spherical harmonics basis functions. The real-time operations in the signal chain of auralization are reduced to the implementation of the head rotation and the convolution with the HRTFs (which are of short length and thus much less expensive as convolution with the RIR).

The signal processing chain is depicted in Fig. 6.6. An exchange of the receiver directivity enables arbitrary HRTFs (e.g. different artificial heads or individual HRTFs). The remaining operations are simple multiplications of SH coefficients and thus real-time capable even for high spatial resolutions. The HRTFs can be stored in a very compact format by separating the delay and the interaural time-differences from the impulse response, resulting in small filters with only 65 frequency bins [Pel12a]. Represented as a weighted set of spatially continuous SH basis functions no explicit interpolation has to be performed. Even a low order description of the data represents a continuously changing function over varying angles. For a full real-time auralization the number of sources and the length of the impulse response do not impact the computational load. The expensive convolutions for many sources and long room impulse responses are superposed in the pre-processing stage.

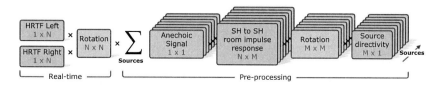

Figure 6.6.: Signal chain to efficiently simulate the transfer paths of several sources and a binaural receiver [Pel12b]

The real-time convolution with directivity patterns represented in terms of their spherical harmonic representation currently allow a maximum order of $n_{\max} \approx 20$ on a standard workstation computer (e.g. Intel Core2 3 GHz) using a buffer size of 256 samples. The auralization can be scaled down for slower machines by only using a subset of the SH coefficients, limiting the maximum order used.

6.3. Summary directivity patterns in room acoustic simulation

With the presented methods arbitrary source and receiver directivity patterns can be implemented in room acoustic simulation algorithms. For particle based methods the implementation can be done in a straightforward manner by applying a complex direction dependent sensitivity. The directivity patterns of musical instrument recordings (cf. Sec. 3) have been implemented in the room simulation software RAVEN [Sch11]. The spherical Fourier transform can be also used to interpolate the directivity patterns onto a rectangular grid as required for the exchange of such data between different implementations [Wef10].

Using a sequential approach it is possible to pre-calculate the room acoustics for a set of sources and transfer paths to a receiver point with arbitrary directivity.

These receiver directivity patterns can then be changed in the post-processing stage. This is in particular advantageous for complex sound scenes using a large number of sources to avoid the need of recalculation of the entire scene for a change in the receiver directivity pattern. This allows the flexible exchange of the receiver directivity patterns on perviously simulated acoustic scenarios, allowing to e.g. implement free head-rotations or the use of individual HRTFs in the post-processing stage after having finished the computationally expensive part of the room simulation.

In wave based simulation methods directivity pattern have been implemented by extending the analytical calculation of rooms as described by Kuttruff [Kut00]. Using the Cartesian derivatives of the eigenfunctions instead of the eigenfunctions itself, the room transfer functions (RTF) can be calculated for the set of physical multipoles (such as monopole, dipole or higher order multipoles) in all possible orientations. These physical multipoles allow to synthesize any arbitrary directivity pattern given as spherical wave spectrum (SWS) using LU decomposition, so the RTF can be found for any pair of arbitrary source and receiver directivity using a superposition approach.

7

Conclusions and Outlook

In this thesis a methodology for the implementation of realistic source and receiver directivity patterns in room acoustical measurements and simulations is described. The procedure is realized in two steps: Firstly, the directivity patterns of the desired sound sources and receivers are obtained and, secondly, the room response for the specific pair of source and receiver directivity patterns is derived using either measurement or simulation. The directivity synthesis can be explicitly calculated for the desired target directivity patterns or implicitly as a superposition result of a suitable set of general directivity patterns.

In order to obtain the source directivity patterns of musical instruments, a set of instruments have been acoustically examined using a surrounding spherical microphone array for simultaneous recording. As the excitation of natural sound sources is not repeatable in a precise manner, these directivity patterns cannot be measured with a sequential measurement method. Due to the relative low spatial resolution of such arrays, spatial undersampling commonly occurs for higher frequencies, especially when the actual sound source is not positioned in the geometric center of the microphone array.

For the analysis of the recorded directivity patterns two methods have been applied: Capturing time-averaged directivity patterns for an instrument and evaluating the complex radiation patterns (including phase relations) obtained individually for all partial tones of all played pitches. The former approach is commonly found in literature [Vig07; Sch11] and employs usually longer recordings of several notes or pieces of music. Spatially and spectrally smooth directivity patterns can be derived that are given as amplitudes or levels without phase relations. Using these averaged directivity data as synthesis target, either a linear approach by choosing an arbitrary phase can be followed or a nonlinear optimization approach that allows to derive arbitrary phase values (and yield results more closer to the target) [Pol10a].

Tonal musical instruments can have different directivity patterns at identical frequencies that are excited from different partial tones of different pitches. The directivity of these tones can be studied after recording the radiation with magnitude and phase information for each played pitch. In order to verify whether the directivity of a natural source is only depending on frequency, the cross-correlation coefficients for all pairs of directivity patterns have been calculated and analyzed. The woodwind instruments show remarkable patterns having frequency blocks of high correlation coefficients spectrally adjacent to blocks of low coefficients. An explanation can be found in the geometry of the woodwinds, causing abrupt changes at some tones in the chromatic scale. Other instrument types behave more similar to purely frequency dependent radiators. The Brass instruments, e.g., show frequency dependent directivity patterns independent of the played pitch that are well suited for obtaining a frequency-dependent average of the directivity.

On the receiver side of the acoustical transfer path, measured or simulated head-related transfer functions (HRTFs) can be used as directivity pattern, being able to deliver a binaural representation for rooms acoustical scenes. From literature on HRTFs it can be concluded that the required spatial resolution is significantly higher on the receiver side as humans are able to localize sound with remarkably high accuracy [Bla97]. These high-resolution HRTFs can be measured using different setups, focusing on either highest possible resolution (for dummy head measurements) or on a considerable fast measurement procedure (for HRTFs measured on human subjects). Alternatively, arbitraty directivity patterns of specific recording microphones can be implemented on the receiver side for the measurement of directivity dependent room impulse responses.

To include the desired directivity patterns in room acoustical measurements these radiation patterns have to be present at the time of the measurement. Commercial products for the process of sound capturing are available, so the focus of this work has been set to develop spherical loudspeaker arrays that can be used for room transfer function (RTF) measurements. The relatively large membrane sizes of the loudspeaker array results in a rather low maximum spherical harmonics order. Using a sequential measurement approach, however, allows to achieve a much higher spatial resolutions. As proof of concept measurements of RIRs have been performed sequentially using a source of known radiation in order to calculate RIRs for arbitrary source directivity patterns. In the implemented example, the measurement duration is only a few minutes, so significant effects from time variances can be disregarded.

A novel measurement device for this method has been designed recently by Klein [Kle12a], which allows to synthesize directivity patterns of SH orders up to 23 using sequential measurements, as long as the requirements on the LTI behavior of the whole system are fulfilled. While simulated room impulse responses can be derived with almost arbitrary high resolution, measurement results are more prone to errors due to possible changes of the physical constants such as air temperature and humidity. Optimized compact spherical loudspeaker arrays are used to obtain room response measurements with respect to directivity patterns of high resolution in a comparably short measurement duration of a few hours for such a massive number of RIR measurements.

Directivity patterns of sources and receiver can also be applied in room acoustical simulation software. In this work methods have been derived to include arbitrary directivity patterns in both wave based simulations and particle based simulations. By using a pre-calculation of the room responses for a set of general directivity patterns, the individual directivity patterns can be derived computationally efficient by superposition in the post-processing stage.

An outlook for the future is certainly the measurement of RTFs with directivity patterns implemented on both sender and receiver side, using a combination of an optimized spherical sound source and a spherical microphone array. As these measurement usually take a considerable amount of time, the trade-offs between time-invariance of the room and obtainable spatial resolution have to be discussed and psychoacoustically evaluated. Employing a block sparsity approach might help to find well suited data for the synthesis that has been recorded within a relative small time frame, so that time-variances can be minimized. This can be done by using only a subset of the measurement results depending on the frequency range and this the required spatial resolution. In future also listening tests have to be employed to derive guidelines for the required spatial resolution of both source and receiver directivity patterns and for the required degree of accuracy of the RTFs with respect to the directivity patterns.

The derived theoretic framework is useful for a wide range of acoustical applications. As the presented methodology separates the implementation of the source, the receiver, and the simulated room, clear interfaces exist to be able to exchange the components individually. These interfaces have the potential to allow the combination of different measurement and simulation results (of source directivity, receiver directivity, or the resulting room response), to compare the results from different scientific institutions for any of these measures. Future research using psychoacoustics will allow to judge the subjective significance for humans regarding the additional parameters of source and receiver directivity in room acoustical measurements and simulations.

A

Appendix

A.1. Geometry of the used measurement devices

In this section the used measurement devices are depicted and briefly described. For the fixed arrays the geometry information is given in a table.

A.1.1. Surrounding flexible array (turntable and arm)

The computerized *arm and turntable* measurement system allows flexible positioning of both azimuth and elevation at the upper part of a full sphere. The elevation angles are limited to approx. $\theta_{\mathrm{max}} = 120°$ measured at a radius of approx. 1.75 m – 2.1 m, depending on the revision of the arm as used over the recent years. Several optimizations of the measurement setup as depicted have been performed since the first construction in 2001. In the current stage of its evolution the trellis arm construction is made of highly stiff carbon material. In comparison to the old, slightly flexible trellis structure made from brass this construction offers much higher positioning accuracy.

Usually rectangular samplings (e.g. the Gaussian sampling or an equiangular sampling) are used for the positions as they allow slightly faster approaching of the subsequent measurement points. The system can be used to measure both source and receiver directivity, attaching a microphone (cf. Fig. 5.8) or a loudspeaker (cf. Fig. 4.2) at the end of the arm. The sequential arm and turntable measurement system can thus be used for the measurement of HRTFs from artificial heads as described in Sec. 4.1.2 or directivity patterns of technical sound sources shown in Sec. 5.3.3.

A.1.2. Surrounding spherical LS array (HRTF arc)

The lightweight *HRTF arc* measurement system has been designed by Masiero [Mas12] and is optimized for the fast measurement of individual HRTFs. It consists of 40 loudspeakers arranged in a circular arc, for details see Sec. 4.1.3.

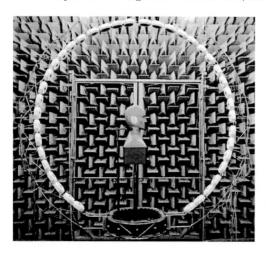

Ch.Nr.	r	θ	ϕ	Ch.Nr.	r	θ	ϕ
20	1 m	6.52°	0°	21	1 m	2.84°	180°
19	1 m	13.93°	0°	22	1 m	10.22°	180°
18	1 m	21.35°	0°	23	1 m	17.64°	180°
17	1 m	28.77°	0°	24	1 m	25.06°	180°
16	1 m	36.19°	0°	25	1 m	32.48°	180°
15	1 m	43.61°	0°	26	1 m	39.90°	180°
14	1 m	51.03°	0°	27	1 m	47.32°	180°
13	1 m	58.46°	0°	28	1 m	54.74°	180°
12	1 m	65.88°	0°	29	1 m	62.17°	180°
11	1 m	73.30°	0°	30	1 m	69.59°	180°
10	1 m	80.72°	0°	31	1 m	77.01°	180°
09	1 m	88.14°	0°	32	1 m	84.43°	180°
08	1 m	95.57°	0°	33	1 m	91.86°	180°
07	1 m	102.99°	0°	34	1 m	99.28°	180°
06	1 m	110.41°	0°	35	1 m	106.70°	180°
05	1 m	117.83°	0°	36	1 m	114.12°	180°
04	1 m	125.26°	0°	37	1 m	121.54°	180°
03	1 m	132.68°	0°	38	1 m	128.97°	180°
02	1 m	140.10°	0°	40	1 m	143.81°	180°
01	1 m	147.52°	0°	39	1 m	136.39°	180°

A.1.3. Surrounding spherical microphone array (24 channels)

The surrounding microphone array has been used by Lentz [Len07] to derive directivity patterns of musical instruments using an energetic averaging approach. The resulting (magnitude only) data is used for the directivity synthesis using a dodecahedron loudspeaker in Sec. 5.4.1.

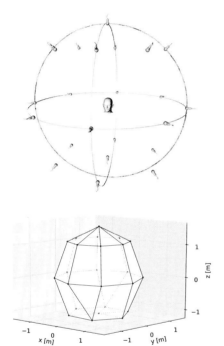

Ch.Nr.	r	θ	ϕ
01	1.56 m	0°	0°
02	1.56 m	45°	0°
03	1.56 m	45°	45°
04	1.56 m	45°	90°
05	1.56 m	45°	135°
06	1.56 m	45°	180°
07	1.56 m	45°	225°
08	1.56 m	45°	270°
09	1.56 m	45°	315°
10	1.56 m	90°	0°
11	1.56 m	90°	45°
12	1.56 m	90°	90°
13	1.56 m	90°	135°
14	1.56 m	90°	180°
15	1.56 m	90°	225°
16	1.56 m	90°	270°
17	1.56 m	90°	315°
18	1.56 m	135°	0°
19	1.56 m	135°	60°
20	1.56 m	135°	120°
21	1.56 m	135°	180°
22	1.56 m	135°	240°
23	1.56 m	135°	300°
24	1.1 m	180°	0°

A.1.4. Surrounding spherical microphone array (32 channels)

The large surrounding microphone array has been designed for accurate measurements of musical instrument directivity patterns, cf. Sec. 3.1.3 and Behler et al. [Beh08]. The calibration of the channels is described in Appendix A.2.1, the geometry of the single speakers has been improved to minimize the impact of the other objects as stated in Appendix A.2.2.

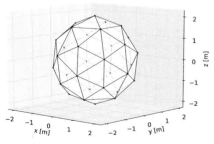

Mic.Nr.[a]	r	θ	ϕ
01 (01)	2.06 m	37.38°	36°
02 (09)	2.06 m	37.38°	108°
03 (17)	2.06 m	37.38°	180°
04 (18)	2.06 m	37.38°	252°
05 (25)	2.06 m	37.38°	324°
06 (02)	2.06 m	79.19°	36°
07 (10)	2.06 m	79.19°	108°
08 (19)	2.06 m	79.19°	180°
09 (20)	2.06 m	79.19°	252°
10 (26)	2.06 m	79.19°	324°
11 (03)	2.06 m	100.81°	72°
12 (11)	2.06 m	100.81°	144°
13 (21)	2.06 m	100.81°	216°
14 (27)	2.06 m	100.81°	288°
15 (04)	2.06 m	100.81°	0°
16 (05)	2.06 m	142.62°	72°
17 (12)	2.06 m	142.62°	144°
18 (22)	2.06 m	142.62°	216°
19 (28)	2.06 m	142.62°	288°
20 (06)	2.06 m	142.62°	0°
21 (13)	2.085 m	0°	0°
22 (29)	2.085 m	63.43°	0°
23 (07)	2.085 m	63.43°	72°
24 (14)	2.085 m	63.43°	144°
25 (23)	2.085 m	63.43°	216°
26 (30)	2.085 m	63.43°	288°
27 (15)	2.085 m	116.57°	108°
28 (24)	2.085 m	116.57°	180°
29 (31)	2.085 m	116.57°	252°
30 (32)	2.085 m	116.57°	324°
31 (08)	2.085 m	116.57°	36°
32 (16)	2.085 m	180°	0°

[a]The channel numbers of the recording setup are given in brackets

A.1.5. Spherical loudspeaker array (12 channels, midrange device)

The spherical loudspeaker array in shape of a dodecahedron has been constructed to allow multichannel excitation. In the following table the data of the mid-range loudspeaker is given. For higher frequencies a 12 channel tweeter dodecahedron with a diameter of 10 cm (instead of 30 cm) is used. The dodecahedron array is used for the synthesis of directivity patterns, cf. Sec. 5.3.1 and Sec. 5.4.1.

Ch.Nr.	r_{mem}	r	θ	ϕ
01	5.2 cm	15 cm	37.38°	0°
02	5.2 cm	15 cm	37.38°	120°
03	5.2 cm	15 cm	37.38°	245°
04	5.2 cm	15 cm	79.19°	60°
05	5.2 cm	15 cm	79.19°	180°
06	5.2 cm	15 cm	79.19°	300°
07	5.2 cm	15 cm	100.81°	0°
08	5.2 cm	15 cm	100.81°	120°
09	5.2 cm	15 cm	100.81°	240°
10	5.2 cm	15 cm	142.62°	60°
11	5.2 cm	15 cm	142.62°	180°
12	5.2 cm	15 cm	142.62°	300°

For a better distribution of the loudspeakers the dodecahedron device as shown above was tilted by 38.5° around the z-axis and 16.2° around the y-axis, resulting in a modified set of membrane positions. This array system is used in Sec. 5.4.2 for the synthesis of room impulse responses for arbitraty source directivity.

Ch.Nr.	r_{mem}	r	θ	ϕ
01	5.2 cm	15 cm	50.91°	29.14°
02	5.2 cm	15 cm	22.97°	145.25°
03	5.2 cm	15 cm	42.43°	297.15°
04	5.2 cm	15 cm	77.25°	95.12°
05	5.2 cm	15 cm	66.76°	221.72°
06	5.2 cm	15 cm	94.29°	338.84°
07	5.2 cm	15 cm	113.24°	41.72°
08	5.2 cm	15 cm	85.71°	158.84°
09	5.2 cm	15 cm	102.75°	275.12°
10	5.2 cm	15 cm	137.57°	117.15°
11	5.2 cm	15 cm	129.09°	209.14°
12	5.2 cm	15 cm	157.03°	325.25°

A.1.6. Spherical loudspeaker array (28 channels, SLAYER)

The spherical loudspeaker array $SLAYER^1$ has been designed by Klein [Kle12a] in order to facilitate broadband excitation both in temporal spectrum and spherical wave spectrum, cf. Sec. 5.3.2. The measured directivity of this device is used for the synthesis of HRTFs in Sec. 5.4.3. Both source and target directivity patterns have been measured using the arm and turntable system as described in Appendix A.1.1.

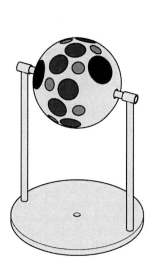

Ch.Nr.	r_{mem}	r	θ	ϕ
01	5.25 cm	20 cm	48.56°	75°
02	5.25 cm	20 cm	70.12°	310°
03	5.25 cm	20 cm	109.88°	235°
04	5.25 cm	20 cm	131.44°	105°
05	3.2 cm	20 cm	11.02°	15°
06	3.2 cm	20 cm	25.30°	180°
07	3.2 cm	20 cm	39.65°	345°
08	3.2 cm	20 cm	54.03°	160°
09	3.2 cm	20 cm	68.42°	15°
10	3.2 cm	20 cm	82.81°	180°
11	3.2 cm	20 cm	97.19°	0°
12	3.2 cm	20 cm	111.58°	165°
13	3.2 cm	20 cm	125.97°	17°
14	3.2 cm	20 cm	140.35°	195°
15	3.2 cm	20 cm	154.70°	330°
16	3.2 cm	20 cm	168.98°	120°
17	2.2 cm	20 cm	21.02°	252°
18	2.2 cm	20 cm	25.30°	110°
19	2.2 cm	20 cm	39.65°	30°
20	2.2 cm	20 cm	54.03°	195°
21	2.2 cm	20 cm	68.42°	345°
22	2.2 cm	20 cm	82.81°	150°
23	2.2 cm	20 cm	97.19°	30°
24	2.2 cm	20 cm	111.58°	195°
25	2.2 cm	20 cm	125.97°	345°
26	2.2 cm	20 cm	140.35°	150°
27	2.2 cm	20 cm	154.70°	33°
28	2.2 cm	20 cm	160.98°	55°

---

[1] Spherical Loudspeaker Array Yielding Exceptional Results

A.2. Implementation of array measurements

In this section the calibration of a spherical microphone array as used for the measurements of tone recordings is described. Improvements of the design for the surrounding array are described, showing an analysis of the individual sensor directivity patterns.

A.2.1. Calibration of surrounding spherical microphone array

A necessary prerequisite of acoustical measurements is the calibration of the whole measurement chain. For the mere use of determining the directivity patterns an absolute calibration is not necessarily required. Nevertheless, fully calibrated data allows to obtain information of the exact levels that are recorded.

Using the concept of the calibrated measurement chain formulated by Dietrich [Die13] the calibration procedures can be derived. As different input sensitivities might be required for compensating the differing sound power of different musical instruments, the possibility of gain adjustments is crucial in order to avoid clipping artifacts or a too small signal-to-noise ratio for the recording of high amplitudes or small amplitudes, respectively.

Before the recording the adjustable gain of the used preamplifiers have been adjusted to sensible positions. During the measurement session these levels have constantly been monitored in order to avoid both noise amplification and clipping. For loud instruments such as the tuba the levels had to be lowered to avoid clipping artifacts, for subsequent low volume instuments the sensitivity had to be increased afterwards. After each of these changes a generated measurement signal has been applied to all of the 32 input channels using a shortcut from the defined output to the input channels.[2] The recorded signal gives the sensitivity with respect to the defined output signal and yields 32 individual sensitivity values that are independent of frequency and quantify precisely the gain settings of the pre-amplifier, used later for the total calibration of the system.

The calibration has been finalized with the measurement of one of the most accessible measurement sensor. The microphone on the south pole of the array has been chosen, as it allows the use of the pistonphone in a stable position as of its vertical alignment. This value allows to calculate the absolute sensitivity from the relative sensitivities determined before.

---

[2]The measurement has been repeated four times as the break-out cable could feed 8 of the 32 channels.

A.2.2. Optimizing the directivity of the array sensors

The sensors of the measurement arrays used have specific directivity patterns, yielding a directional dependent sensitivity. In order to avoid large deviations from the angle of incidence, a rather omni-directional directivity is beneficial, at least in the directions of interest. An optimization has been performed for the surrounding microphone array (cf. Appendix A.1.1) and the surrounding loudspeaker array (cf. Appendix A.1.2).

Sensors of surrounding microphone array

In Fig. A.1a the maximum deviation of all sensitivity values within a certain cone of angles around the on-axis result are plotted. The uncertainties caused by the variance of the directivity can be quantified with this plot. At some frequencies such as 8 and 11 kHz deviations of up to 10 dB can be observed if the array is used without additional absorbers. In this section the modifications of the microphone array are stated explicitly, yielding much smoother directivity patterns for the microphone elements as visible in Fig. A.1b.

Optimization of surrounding spherical microphone array

The calibration shows a dependence on the angle of sound incidence. In order to separate the influence of the housing of the microphones, a reciprocal BEM simulation has been performed. The geometry of the microphone in the mounting situation is simulated with a point source at the opening of the microphone membrane, allowing to calculate the transfer path from any point in space to the microphone. In Fig. A.2 the logarithmic directivity value of a single microphone is plotted for the angles of sound incidence of $0°$, $20°$ and $40°$ with respect to the on-axis radiation. The on-axis sensitivity shows a remarkable variation, much bigger than the commonly used rules-of-thumb suggest. The artifacts begin already at rather low frequencies of approx. 2 kHz, where the objects are still small compared to the wavelength. As the fiber glass sticks and the connectors have diameters from up to 1 cm, this influence is expected to be caused by the high symmetry of the stick geometry. For a different angle of sound incidence, the comb filter effect visible in Fig. A.2 is reduced.

In order to confirm the BEM simulation, the array microphone has been measured with and without suitable modifications to increase the smoothness of their directional sensitivity. The measurement data is obtained for angles up to $45°$

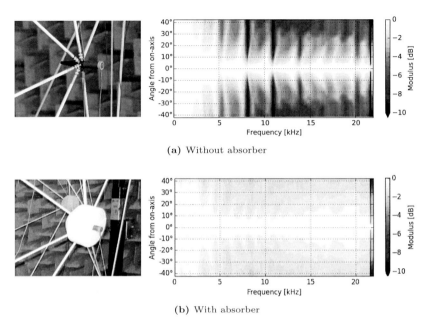

(a) Without absorber

(b) With absorber

Figure A.1.: Maximum SPL deviation from the on-axis response of an array microphone with and without baffle absorber.

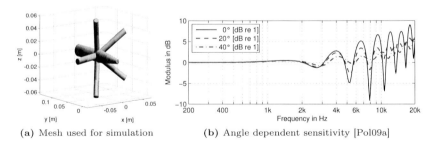

(a) Mesh used for simulation

(b) Angle dependent sensitivity [Pol09a]

Figure A.2.: Boundary-Element-Method simulation of the angle dependent microphone sensitivities as used in the surrounding spherical microphone array

from the on-axis direction, corresponding roughly to a maximum deviation of 1.5 m from the geometric center of the array as used for the recording of musical instruments.

In Fig. A.3 the logarithmic directivity for a set of angular deviation from the on-axis response of the original loudspeaker and one of the test version are depicted, showing a much smoother sensitivity especially in the on-axis direction. The smooth frequency response is achieved by the use of a foam absorber that can be attached to the microphone housing. Several modifications of the microphone housing have been tested (e.g. a drop-shaped solid chassis and several foam absorbers of different shapes and sizes using *Polyurethane* (PU) foam and *Basotect*[3] foam). The depicted absorber consists of an attachable cylindrical block of Basotect with grated edges, as depicted in Fig. A.4.

Directivity analysis of loudspeaker array

The original setup as developed by Masiero [Mas12] and an improved version using a foam absorber made of *Basotect* are depicted in Fig. A.5.

The directivity pattern of the loudspeaker consists of its original directivity and the reflections from the physical objects in vicinity of the loudspeaker. While the influence of objects located at a distance of greater than approx. one meter from the loudspeaker can be eliminated applying a suitable time window, the reflections at the immediate neighbors of the actual speaker have to be minimized.

In Fig. A.6a the part of the loudspeaker arc from both setups is analyzed with and without neighboring loudspeakers. The setup is depicted on the left, while the resulting deviations from the on-axis radiation are depicted on the right. For these plots the directivity patterns have been measured on a Gaussian sampling of order $n_{max} = 35$ limited to the range of $\theta \leq 22°$ with the axis of the loudspeaker pointing to the positive z-axis. The highest deviation of the measured levels between on-axis result and all points with a maximum angular deviation are plotted in the contour plot on the right. The color is segmented in intervals of 0.5 dB deviation from the on-axis result. It can be seen that for the original setup the maximum deviation is approx. 4 dB up to frequencies of around 15 kHz, if assuming 14° as the required angular limit for even sound distribution, cf. Fig. 4.4.

---

[3]Basotect is an optimized foam material for the use as acoustic absorber. Further information can be found at http://www.basf.com/group/corporate/de/brand/BASOTECT

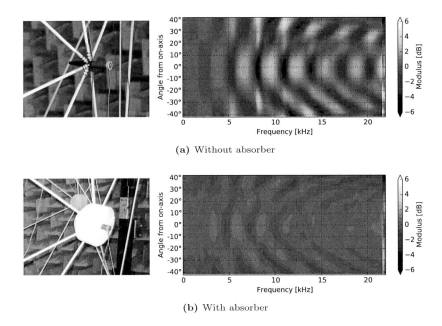

(a) Without absorber

(b) With absorber

Figure A.3.: Directivity value of array microphone with and without baffle absorber.

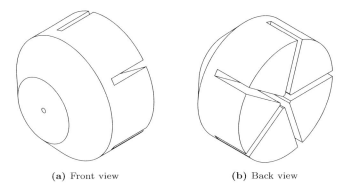

(a) Front view　　　　　　　(b) Back view

Figure A.4.: Absorber for obtaining a smooth directivity pattern for the sensors of the microphone array. Here the version for the 5-stick element is depicted.

(a) Original design   (b) Improved design

Figure A.5.: Original transducer design used and improved version using a foam absorber as used in the measurement arc for fast HRTF measurements.

The directivity pattern of the same loudspeaker without its neighbors is depicted in Fig. A.6b. The influence on the sound distribution onto the area of interest is enhanced significantly, with deviations for less than 1 dB up to frequencies of approx. 10 kHz. In the higher frequency rage the presence of the other transducers does not change the encountered levels, suggesting that the directivity of a single transducer is already so strongly focused that the interference due to the neighboring transducer chassis is not significant anymore.

As the fast measurement method of HRTFs cannot be performed without using multiple speakers, the frequency response is optimized by using acoustical absorbers around the loudspeaker chassis.

In Fig. A.6c the single loudspeaker response is depicted for the same speaker with an acoustic absorber attached. The contour plot shows very similar directivity patterns as the single speaker without any absorber. Only the resonances seem to be a little bit more smooth. Furthermore a noticeable boost for low frequencies could be discovered when using the baffles made from absorber foam. As the HRTFs are referenced, this frequency dependent amplitude variation only has a minor effect, but enhances the SNR in low frequency marginally.

Lastly the effect of the absorbers is tested on the HRTF arc as used for measurements, depicted in Fig. A.6d. While the negative influence of the loudspeaker is still visible, a clear enhancement can be achieved using the absorbing foam. Especially the improvement in the frequency range between 2 kHz and 8 kHz is expected to be highly relevant, as for a correct interference pattern the intensity

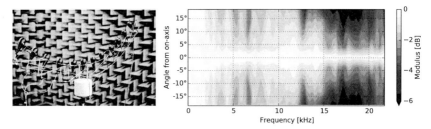

(a) 12 drop shaped loudspeaker chassis mounted on the HRTF measurement arc

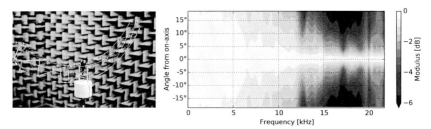

(b) Single drop shaped loudspeaker chassis mounted on the HRTF measurement arc

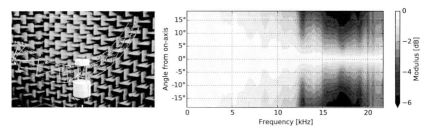

(c) Single loudspeaker mounted in absorber on the HRTF measurement arc

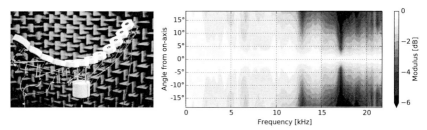

(d) 12 loudspeakers mounted in absorbers on the HRTF measurement arc

Figure A.6.: Maximum SPL deviation from the on-axis response of a single loudspeaker in different mounting setups.

of the wave arriving at the ear cannel with shoulder reflection and using the direct path has to be identical.

A.3. Spherical datasets

A.3.1. Musical instrument recordings (Berlin, 2009)

The following modern instruments have been measured in a collaboration with the Audio Communication Group of TU Berlin [Beh12]:

- **Woodwinds** (Bassoon, Contrabassoon, Oboe, English horn, Clarinet, Bass clarinet, Alto saxophone, Tenor saxophone, Western concert Boehm flute)

- **Brass instruments** (Trumpet, Tenor trombone, Bass trombone, Tenor saxophone, Tuba, French horn)

- **Percussion** (Timpani)

- **Singer** (Soprano)

- **String instruments** (Viola, Violin, Cello, Double-action pedal harp, Contrabass, Classical guitar)

From the group of historical instruments additionally these instruments have been measured:

- **Woodwinds** (Basset horn, Bassoon, Western concert Boehm flute, Clarinet, Oboe)

- **Brass instruments** (Bass trombone, French horn, Tenor trombone, Trumpet)

- **String instruments** (Contrabass, Violin, Viola, Cello)

The modern instruments have been played mostly by members of the *German Symphony Orchestra Berlin* joined by few other professional orchestras. The historical instruments have been played by members of *Akademie für Alte Musik Berlin* [Det10; Krä10]. The microphone array has been used without the enhancements using a specialized absorber as described in Appendix A.2.2.

A.3.2. HRTFs measurement data at different ranges

The measurement data used for the range extrapolation has been obtained at radii of 200 cm and a set of closer distances, measured from Lentz [Len07] in 2001 and in 2004, respectively. The setup most likely has been modified and a direct comparison of the measurement data does not show comparable levels of the HRTFs. Even the data obtained on different ranges in one measurement session from 2004 shows some gain variances in the data, presumably to make best use of the available headroom. In order to resolve this issue the DTF as mentioned by Middlebrooks [Mid99] have been calculated. Instead of the correlation coefficient, the normalized correlation coefficients are used, allowing to perform the analysis of the data without the need of performing extensive new measurement sessions.

A.3.3. SLAYER Radiation

The results of the directivity measurements of the optimized spherical loud-speaker array SLAYER device (cf. Appendix A.1.6) are visualized here exemplarily for one of each membrane types. The measurement has been performed using the computerized positioning system as described in Appendix A.1.1 and depiced in Fig. 5.8. A Gaussian quadrature sampling scheme of order $n_{\max} = 82$ as described in Sec. 2.5.2 has been used as measurement grid.

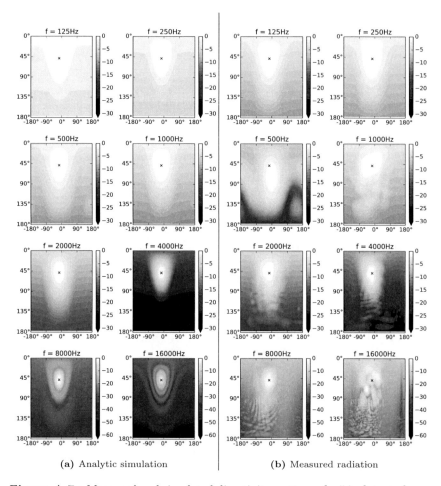

(a) Analytic simulation　　　　　　　(b) Measured radiation

Figure A.7.: Measured and simulated directivity pattern of a 5 inch transducer mounted on a sphere. Center of membrane marked with a black cross.

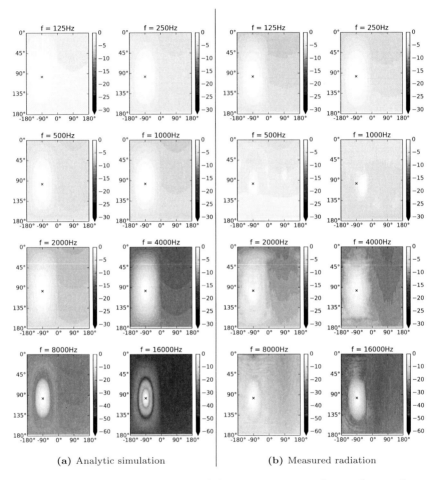

(a) Analytic simulation   (b) Measured radiation

Figure A.8.: Measured and simulated directivity pattern of a 3 inch transducer mounted on a sphere. Center of membrane marked with a black cross.

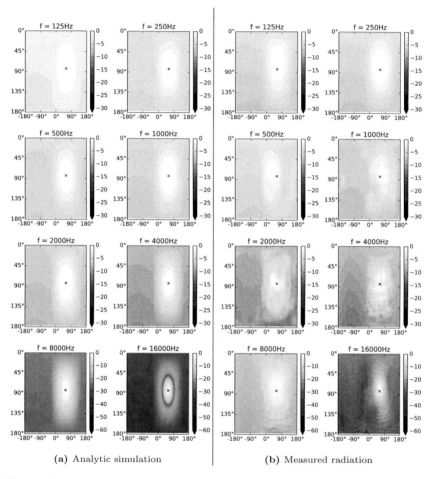

(a) Analytic simulation   (b) Measured radiation

Figure A.9.: Measured and simulated directivity pattern of a 2 inch transducer mounted on a sphere. Center of membrane marked with a black cross.

A.4. Software tools, data processing and visualization

A.4.1. ITA-Toolbox for Matlab

All acoustical measurements used for this thesis have been performed using the *ITA-Toolbox* for Matlab, an open source project[4] covering a broad spectrum of useful routines for acoustic engineers. The ITA-Toolbox project stated as an initiative to develop a common code base for the employees of the Institute of Technical Acoustics (ITA) of RWTH Aachen University. The project quickly became more robust and more diverse, so that for most of the fundamental functions used for audio and acoustic engineering well tested, stable solutions exist. Nowadays the ITA-Toolbox can be regarded as standard software toolkit at ITA. It allows the implementation of the full signal chain, including measurements using standard sound cards and various post-processing routines using Matlab. In scope of the work presented in this thesis a package that implements the fundamental concepts of processing in the spherical harmonic domain has been added to ITA-Toolbox. It implements the calculation of e.g. the spherical harmonics, spherical Hankel and Bessel functions, plotting functions to visualize data in the spatial or Fourier domain and the sampling schemes of all array systems used for this thesis.

A.4.2. Data formats (HDF5, OpenDAFF)

For very high resolution data the usual way of storing data from the ITA-Toolbox results in a very large number of files allowing fast access only when retrieving the data for a single channel. As an alternative for fast random access the *Hierarchical Data Format* (HDF) has been used to store complete datasets in single files. Its current version (HDF5) provides an open-source file format and library for storing and managing data in large files. It is "designed for flexible and efficient I/O and for high volume and complex data"[5]. The new SOFA format for storing spatial audio data also employs HDF5 as fundamental data container [Maj13b]. As an alternative, OpenDAFF can be employed as described in Sec. 2.5.4, as long as the requirements on the used sampling scheme are fulfilled [Wef10].[6]

---

[4]http://www.ita-toolbox.org
[5]http://www.hdfgroup.org/HDF5
[6]The current version of OpenDAFF supports only rectangular equiangular sampling schemes, as described in Sec. 2.5.1 and Sec. 2.5.4.

A.4.3. Implementation of Balloon plots

For the visualization of directivity patterns many plot types can be employed and prove useful for specific applications. A very descriptive visualization (especially for interactive applications) is the balloon plot, depicting the magnitude of a spherical function as the radius of a continuous balloon plot. An additional dimension of the data can be visualized using the color of the surface of this balloon. Usually spherical measurement results of a physical quantity can be plotted using the following values [Pol14b].

The magnitude M, the sound pressure level L and the angular phase A for a directivity described as spherical pressure function $p(\theta, \phi)$ are calculated:

$$M = |p(\theta, \phi)| \tag{A.1}$$

$$L = 20 \log_{10} \left(\frac{|p(\theta, \phi)|}{\max (|p(\theta, \phi)|)} \right) \text{dB} + \Delta L \tag{A.2}$$

$$A = \arg (p(\theta, \phi)) \tag{A.3}$$

As the logarithmic scale is not bound, an effective dynamic range ΔL is chosen as the maximum level and all values below $0 \, \text{dB}$ are truncated to a radius of zero. Various plot types can be composed from these physical quantities by a variation of the radius r and the color c, with the following plots supported by the developed software:

- Complex balloon plot ($r = M$, $c = A$)

- Log. complex balloon plot ($r = L$, $c = A$)

- Magnitude balloon plot ($r = c = M$)

- Log. magnitude balloon plot ($r = c = L$)

- Magnitude on unit sphere ($r = 1$, $c = M$)

- Log. magnitude on unit sphere ($r = 1$, $c = L$)

- Phase on unit sphere ($r = 1$, $c = A$)

These plot types allow an efficient graphical analysis of spherical data. The used colormaps can be adjusted to express either a linear scale (such as amplitude) or a circular scale (such as phase) that have identical colors for the starting and ending point of the colormap.

Balloon plots of a rectangular sampling scheme can be implemented by the calculation of small polygons connecting four vertices. A more general representation that also works for irregular sampling schemes is implemented by triangularization. Hereby the intersection of all discrete angles with the unit sphere is defined as points in the 3D space. Then a triangularization of the data is performed by the computation of the convex hull covering all sampling points in order to get a complete closed balloon.

A.4.4. Visual quality assessment of spherical data

Working with large sets of scientific data is challenging in terms of how to obtain a general understanding of the data to be able to verify the accuracy and quality of the obtained results. Visual evaluation of large sets of measurement data requires tools that can be employed. Even if mathematical methods are expected to yield more precise results, a visual check can be beneficial to rapidly check the data of e.g. acoustical measurements and trace down some of the systematic or unsystematic errors.

The visualization package has been implemented in Python using the open source package VTK via MayaVi as API. The data is read using the package *h5py* which allows fast access of data stored in the HDF5 format. Details of the implementation can be found in Pollow et al. [Pol14b]. The result is a graphical user interface as depicted in Fig. A.10 to load and analyze sets of directivity data.

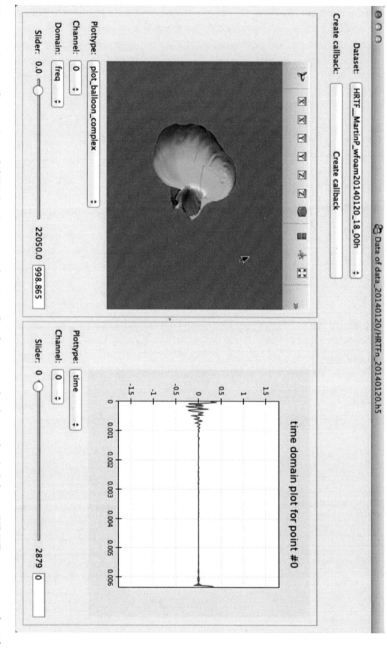

Figure A.10.: Screenshot of GUI for data visualization. In this example the author's left ear HRTF as measured with the setup described in Appendix A.1.2 is plotted as complex balloon plot at a frequency of 1 kHz.

Glossary

List of Acronyms

BEM	Boundary Element Method
CAVE	Cave Automatic Virtual Environment
CTC	Cross-Talk Cancellation
DC	Direct Current
DFT	Discrete Fourier Transform
DSHT	Discrete Spherical Harmonic Transform
DTF	Directional Transfer Function
FEM	Finite Element Method
FFM	Fast Multipole Method
FFT	Fast Fourier Transform
FT	Fourier Transform
FTDS	Fourier Transform of Discrete time Signals
HDF5	Hierarchical Data Format (version 5)
HRIR	Head-Related Impulse Response
HRTF	Head-Related Transfer Function
ILD	Interaural Level Differences
ISHT	Inverse Spherical Harmonic Transform
ISO	International Organization for Standardization
ITA	Institute of Technical Acoustics (RWTH Aachen University)
ITD	Interaural Time Differences
LMS	Least Mean Squares
LTI	Linear Time-Invariant
MESM	Multiple Exponential Sweep Method
RAVEN	Room Acoustics for Virtual Environments
RIR	Room Impulse Response

RTF	Room Transfer Function
SH	Spherical Harmonics
SHT	Spherical Harmonic Transform
SLAYER	Spherical Loudspeaker Array Yielding Exceptional Results
SNR	Signal-to-Noise Ratio
SPL	Sound Pressure Level
SWS	Spherical Wave Spectrum

Mathematical Operators

$\mathcal{F}\{\cdot\}$	Fourier transform
$\mathcal{F}^{-1}\{\cdot\}$	inverse Fourier transform
$\mathcal{S}\{\cdot\}$	spherical harmonic transform
$\mathcal{S}^{-1}\{\cdot\}$	inverse spherical harmonic transform
$\overline{(\cdot)}$	complex conjugate
$(\cdot)^{H}$	Hermitian transpose of matrix
$(\cdot)^{-1}$	inverse of matrix
$(\cdot)^{+}$	Moore-Penrose pseudoinverse of matrix
$\langle\cdot\,|\,\cdot\rangle$	inner product of two functions
$\langle\cdot\rangle_{(\theta',\phi')}$	arithmetic mean over the unit sphere
$|\cdot| = \|\cdot\|_2$	2-norm (Euklidian length) of vector
$\mathrm{diag}\{\cdot\}$	operator that converts a vector to a diagonal matrix
$\mathrm{vec}\{\cdot\}$	operator converting spatially samples values to a vector
$\mathrm{vec_{SH}}\{\cdot\}$	operator converting a spherical wave spectrum to a vector

Mathematical Symbols

(α, β, γ)	Euler angles
(r, θ, ϕ)	spherical coordinate system
α_i	weight of spatial sampling point i
$a_{nm} = a_n$	spherical wave spectrum of aperture function

b_{nm}	multipole expansion coefficients for interior problems
c_{nm}	multipole expansion coefficients for exterior problems
$C(\tau)$	cross-correlation function as function of time lag τ
$C(f, g)$	cross-correlation coefficient of spherical functions f and g
$C(f, g)$	normalized cross-correlation coefficient
$D_0(\theta, \phi)$	directivity factor (referenced to specific direction)
$D_{0,\text{magn}}(\theta, \phi)$	directivity factor of magnitude values
$D_{Lm}(\theta, \phi)$	directivity (logarithmic)
$D_m(\theta, \phi)$	directivity value (referenced to equivalent monopole radiation)
$d_{k,m}^{(n)}(\beta)$	reduced Wigner-D function
$D_{k,m}^{(n)}(\alpha, \beta, \gamma)$	Wigner-D function
$\delta^{(\theta', \phi')}(\theta, \phi)$	Dirac impulse pointing in direction (θ, ϕ) given (spatial domain)
$\delta_{nm}^{(\theta', \phi')}$	Dirac impulse pointing in direction (θ, ϕ) given as SWS
δ_{xy}	Kronecker delta function
$E_a(n)$	geometically summed squared magnitudes
ε	regularization constant
f	frequency
f_N	Nyquist frequency
f_s	sampling rate
$f(t)$	signal in time domain
$f(\theta, \phi)$	spherical function
f_{nm}	spherical wave spectrum (SWS) of spherical function $f(\theta, \phi)$
$\hat{f}(\omega)$	signal in frequency domain
$\hat{\mathbf{f}}$	spherical wave spectrum stored as vector
$\mathbf{f}$	spatially sampled points of a spherical function stored as vector
$G(\mathbf{r}\|\mathbf{r}')$	Green's function for source at $\mathbf{r}'$ observed at position $\mathbf{r}$
$H_{\text{free-field}}^{[\text{L/R}]}(\theta, \phi)$	Head related transfer function
$H(\omega)$	room transfer function
$h_n^{(1)}(\cdot)$	spherical Hankel functions of first kind and order n
$h_n^{(2)}(\cdot) = h_n(\cdot)$	spherical Hankel functions of second kind and order n
$\mathbf{I}$	identity matrix
$j_n(\cdot)$	spherical Bessel function of order n

κ	condition number
$n_{\max}$	maximum order of spherical wave spectrum
$\mathbf{O}$	orthogonality matrix
ω	angular frequency
Ω	room angle
$p(\cdot)$	sound pressure in time or frequency domain
$p_{nm}(r)$	SWS of sound pressure
$\hat{\mathbf{p}}$	SWS of sound pressure stored as vector
$\hat{\mathbf{p}}_{\mathrm{t}}$	SWS of target pressure function stored as vector
$\mathbf{p}_{\mathrm{mic}}$	sound pressure at microphone array positions stored as vector
$\mathbf{p}_{\mathrm{re}}(\mathbf{v})$	reconstructed sound pressure at microphone array positions stored as vector
$\psi_i(\mathbf{r})$	spatially sampled eigenfunctions of rectangular room
$\hat{\mathbf{P}}$	SWS of sound pressure stored as matrix
$P_n(\cdot)$	Legendre polynomials of order n
$P_n^m(\cdot)$	associated Legendre functions of first kind and order n
$Q_n^m(\cdot)$	associated Legendre functions of second kind and order n
$r'(\theta, \phi)$	spherical residual function
$\mathbf{r}$	point in space stored as vector
r_{mem}	membrane radius
$\mathbf{r}_{\mathrm{p}}$	sampled spherical residual function stored as vector
$\mathbf{r}_{\mathrm{abs}}$	residual function of absolute function values stored as vector
$\mathbf{R_y}$	rotation matrix for rotations around the y-axis
$\mathbf{R_z}$	rotation matrix for rotations around the z-axis
S^2	surface of the unit sphere
$\mathbf{u}_{\mathrm{t}}$	target feeding voltage for loudspeaker array
v_{nm}	spherical wave spectrum of surface velocity
$\mathbf{v}$	membrane velocities of louspeaker array stored as vector
$Y_n^m(\theta, \phi)$	spherical harmonics of order n and degree m
$\mathbf{Y} = \mathbf{Y}_N$	matrix of sampled spherical harmonics up to order N
Z_{nm}	acoustic impedance as spherical wave spectrum

Bibliography

[All79]   J. B. Allen and D. A. Berkley. "Image method for efficiently simu-
            lating small-room acoustics". In: *Journal of the Acoustical Society of
            America* 65.4 (1979), pp. 943–950.

[Ant09]   C. Antweiler and G. Enzner. "Perfect Sequence LMS for Rapid Ac-
            quisition of Continuous-Azimuth Head Related Impulse Responses".
            In: *IEEE Workshop on Applications of Signal Processing to Audio
            and Acoustics (WASPAA)*. 2009, pp. 281–284.

[Bad05]   R. Bader. *Computational Mechanics of the Classical Guitar*. Springer,
            2005.

[Bau10]   R. Baumgartner and E. Messner. "Auswirkung der Abstrahlcharak-
            teristik auf die Klangfarbe von Querflöten und Saxofonen". Bachelor
            thesis. TU Graz, 2010.

[Beh02]   G. K. Behler. "Uncertainties of Measured Parameters in Room
            Acoustics Caused by the Directivity of Source and/or Receiver". In:
            Proceedings of Forum Acusticum. Sevilla, Spain, 2002.

[Beh08]   G. K. Behler, M. Pollow, and D. Schröder. "Messung und Simula-
            tion von Raumimpulsantworten für eine realistische Auralisation".
            In: *25. Tonmeistertagung, VDT International Convention*. Leipzig,
            Germany, 2008.

[Beh12]   G. K. Behler, M. Pollow, and M. Vorländer. "Measurements of mu-
            sical instruments with surrounding spherical arrays". In: *Proceedings
            of Acoustics 2012*. Nantes, France, 2012.

[Ber54]   L. L. Beranek. *Acoustics*. McGraw-Hill Book Company, Inc., 1954.

[BH11]   I. Ben Hagai, M. Pollow, M. Vorländer, and B. Rafaely. "Acous-
            tic centering of sources measured by surrounding spherical micro-
            phone arrays". In: *Journal of the Acoustical Society of America* 130.4
            (2011), pp. 2003–2015.

[BI03]   A. Ben-Israel and T. N. Greville. *Generalized inverses*. Springer,
            2003.

[Bla74]   J. Blauert. *Räumliches Hören*. S. Hirzel Verlag, 1974.

[Bla97]   J. Blauert. *Spatial hearing: the psychophysics of human sound localization*. MIT press, 1997.

[Bro00]   I. N. Bronštein and K. A. Semendjaev. *Taschenbuch der Mathematik*. Deutsch, 2000.

[Bro13]   S. Brown and D. Sen. "Error Analysis of Spherical Harmonic Soundfield Representations in Terms of Truncation and Aliasing Errors". In: *ICASSP* (2013).

[Bro95]   A. W. Bronkhorst. "Localization of real and virtual sound sources". In: *The Journal of the Acoustical Society of America* 98.5 (1995), pp. 2542–2553.

[Can06]   E. J. Candès, J. Romberg, and T. Tao. "Robust uncertainty principles: Exact signal reconstruction from highly incomplete frequency information". In: *Information Theory, IEEE Transactions on* 52.2 (2006), pp. 489–509.

[Cre81]   L. Cremer. *Physik der Geige*. Stuttgart: S. Hirzel Verlag, 1981.

[Dal93]   B.-I. Dalenbäck, M. Kleiner, and P. Svensson. "Audibility of changes in geometric shape, source directivity, and absorptive treatment-experiments in auralization". In: *Journal of the Audio Engineering Society* 41.11 (1993), pp. 905–913.

[Deb10]   D. Deboy and F. Zotter. "Acoustic center and orientation analysis of sound-radiation recorded with a surrounding spherical microphone array". In: *Proceedings of the 2nd International Symposium on Ambisonics and Spherical Acoustics*. 2010, pp. 1–6.

[Det10]   E. Detzner, F. Schultz, M. Pollow, and S. Weinzierl. "Zur Schallleistung von modernen und historischen Orchesterinstrumenten II: Holz- und Blechblasinstrumente". In: *36th German Annual Conference on Acoustics (DAGA)*. Berlin, Germany, 2010.

[Die12a]   P. Dietrich, M. Guski, M. Pollow, M. Müller-Trapet, B. Masiero, R. Scharrer, and M. Vorländer. "ITA-Toolbox – An Open Source MATLAB Toolbox for Acousticians". In: *38th German Annual Conference on Acoustics (DAGA)*. Darmstadt, Germany, 2012.

[Die12b]   P. Dietrich, B. Masiero, and M. Vorländer. "On the Optimization of the Multiple Exponential Sweep Method". In: *Journal of the Acoustical Society of America* (2012).

[Die13]   P. Dietrich. "Uncertainties in Acoustical Transfer Functions. Modeling, Measurement and Derivation of Parameters". PhD thesis. RWTH Aachen University, 2013.

[Dri94]   J. Driscoll and D. Healy Jr. "Computing Fourier transforms and convolutions on the 2-sphere". In: *Advances in Applied Mathematics* 15.2 (1994), pp. 202–250.

[Dud98]   R. O. Duda and W. L. Martens. "Range dependence of the response of a spherical head model". In: *The Journal of the Acoustical Society of America* 104.5 (1998), pp. 3048–3058.

[Dur04]   R. Duraiswami, D. N. Zotkin, and N. A. Gumerov. "Interpolation and Range Extrapolation of HRTFs". In: *Proceedings the IEEE Conference of Acoustics, Speech, and Signal Processing*. 2004, pp. 46–48.

[Fel08]   J. Fels. "From Children to Adults: How Binaural Cues and Ear Canal Impedances Grow". PhD thesis. RWTH Aachen University, 2008.

[Gol96]   G. Golub and C. Van Loan. *Matrix computations*. Johns Hopkins University Press Baltimore, MD, USA, 1996.

[Gum02a]   N. A. Gumerov and R. Duraiswami. "Computation of scattering from N spheres using multipole reexpansion". In: *Journal of the Acoustical Society of America* 112.6 (2002), pp. 2688–2701.

[Gum02b]   N. A. Gumerov, R. Duraiswami, and Z. Tang. "Numerical study of the influence of the torso on the HRTF". In: *Acoustics, Speech, and Signal Processing (ICASSP), 2002 IEEE International Conference on*. Vol. 2. IEEE. 2002, pp. II–1965.

[Gum03]   N. A. Gumerov and R. Duraiswami. "Recursions for the computation of multipole translation and rotation coefficients for the 3-D helmholtz equation". In: *SIAM J. Sci. Comput.* 25.4 (2003), pp. 1344–1381.

[Gum05]   N. A. Gumerov and R. Duraiswami. *Fast multipole methods for the Helmholtz equation in three dimensions*. Access Online via Elsevier, 2005.

[Hea03]   D. Healy, D. Rockmore, P. Kostelec, and S. Moore. "FFTs for the 2-Sphere-Improvements and Variations". In: *Journal of Fourier Analysis and Applications* 9.4 (2003), pp. 341–385.

[Hoh09]   F. Hohl. "Kugelmikrofonarray zur Abstrahlungsvermessung von Musikinstrumenten". Diploma thesis. University of Music and Performing Arts, 2009.

[Iso]   *ISO 3382. Acoustics – Measurement of Room Acoustic Parameters.* *2009.*

[Kah98]   Y. Kahana, P. A. Nelson, M Petyt, and S. Choi. "Boundary element simulation of HRTFs and sound fields produced by virtual acoustic imaging systems". In: *Audio Engineering Society Convention 105.* Audio Engineering Society. 1998.

[Kat01]   B. F. Katz. "Boundary element method calculation of individual head-related transfer function. I. Rigid model calculation". In: *The Journal of the Acoustical Society of America* 110.5 (2001), pp. 2440–2448.

[Kle12a]   J. Klein. "Optimization of a Method for the Synthesis of Transfer Functions of Variable Sound Source Directivities for Acoustical Measurements". Diploma thesis. RWTH Aachen University, 2012.

[Kle12b]   J. Klein, M. Pollow, and P. Dietrich. "Optimized System for the Synthesis of Optimized System for the Synthesis of Room Impulse Responses of Arbitrary Sound Sources". In: *16th International Student Conference on Electrical Engineering (POSTER).* Prague, Czech Republic, 2012.

[Kle14]   J. Klein, M. Pollow, and M. Vorländer. "Optimized spherical sound source for auralization with arbitrary directivity". In: *EAA Joint Symposium on Auralization and Ambisonics.* Berlin, Germany, 2014.

[Kos03]   P. Kostelec and D. Rockmore. "FFTs on the rotation group". In: *Santa Fe Institute Working Papers Series* (2003).

[Kou12]   G. I. Koutsouris, J. Brunskog, C.-H. Jeong, and F. Jacobsen. "A combination of the acoustic radiosity and the image source method". In: *41st International Congress and Exposition on Noise Control Engineering.* 2012.

[Kre12]   B. Krechel. "Schnelle Messung von individuellen HRTFs mit kon-
           tinuierlichen MIMO-Verfahren". Bachelor thesis. RWTH Aachen
           University, 2012.

[Krä10]   J. Krämer, F. Schultz, M. Pollow, and S. Weinzierl. "Zur Schall-
           lleistung von modernen und historischen Orchesterinstrumenten I:
           Streichinstrumente". In: *36th German Annual Conference on Acous-
           tics (DAGA)*. Berlin, Germany, 2010.

[Kui99]   J. B. Kuipers. *Quaternions and rotation sequences*. Vol. 66. Prince-
           ton university press Princeton, 1999.

[Kun11]   M. Kunkemöller. "Entwicklung eines Analyse- und Synthesever-
           fahrens von mehrkanalig gemessenen Raumimpulsantworten für vari-
           able Quellrichtcharakteristiken". Diploma thesis. RWTH Aachen
           University, 2011.

[Kun12]   M. Kunkemöller, P. Dietrich, M. Pollow, and M. Vorländer. "Syn-
           thesis of Room Impulse Responses for Arbitrary Source Directivities
           using Spherical Harmonic Decomposition". In: *38th German Annual
           Conference on Acoustics (DAGA)*. Darmstadt, Germany, 2012.

[Kut00]   H. Kuttruff. *Room Acoustics*. Spon Press (UK), 2000.

[Len07]   T. Lentz. "Binaural Technology for Virtual Reality". PhD thesis.
           RWTH Aachen University, 2007.

[Li07]   Z. Li and R. Duraiswami. "Flexible and Optimal Design of Spherical
           Microphone Arrays for Beamforming". In: *IEEE Transactions on
           Acoustics, Speech and Language Processing*. Vol. 15. 2. 2007, pp. 702–
           714.

[Maj07]   P. Majdak, P. Balazs, and B. Laback. "Multiple Exponential Sweep
           Method for Fast Measurement of Head-Related Transfer Functions".
           In: *Journal of the Audio Engineering Society* 55.7/8 (2007), pp. 623–
           637.

[Maj13a]   P. Majdak and H. Ziegelwanger. "Continuous-direction model of the
           broadband time-of-arrival in the head-related transfer functions".
           In: *Proceedings of International Congress on Acoustics*. Montreal,
           Québec, Canada, 2013.

[Maj13b]   P. Majdak, Y. Iwaya, T. Carpentier, R. Nicol, M. Parmentier, A. Roginska, Y. Suzuki, K. Watanabe, H. Wierstorf, H. Ziegelwanger, and M. Noisternig. "Spatially Oriented Format for Acoustics: A Data Exchange Format Representing Head-Related Transfer Functions". In: *AES Convention*. 2013.

[Mar85]   A. Marshall and J Meyer. "The directivity and auditory impressions of singers". In: *Acta Acustica united with Acustica* 58.3 (1985), pp. 130–140.

[Mas12]   B. Masiero. "Individualized Binaural Technology. Measurement, Equalization and Perceptual Evaluation". PhD thesis. RWTH Aachen University, 2012.

[Mec08]   F. P. Mechel. *Formulas of Acoustics*. Springer, 2008.

[Mey02]   J. Meyer and G. Elko. "A highly scaleable spherical microphone array based on an orthonormal decomposition of the soundfield1". In: *IEEE International Conference on Acoustics, Speech, and Signal Processing (ICASSP)*. 2002.

[Mey64]   J. Meyer. "Die Richtcharakteristiken von Geigen". In: *Instrumenten-bau-Zeitschrift* 18 (1964), p. 275.

[Mey65a]   J. Meyer. "Die Richtcharakteristiken des Flügels". In: *Das Musikinstrument* 14 (1965), p. 1085.

[Mey65b]   J. Meyer. "Die Richtcharakteristiken von Klarinetten". In: *Das Musikinstrument* 14 (1965), p. 21.

[Mey65c]   J. Meyer. "Die Richtcharakteristiken von Violoncelli". In: *Instrumentenbau-Zeitschrift* 19 (1965), p. 281.

[Mey66a]   J. Meyer. "Akustik der Holzblasinstrumente in Einzeldarstellungen". In: *Das Musikinstrument* 17 (1966).

[Mey66b]   J. Meyer. "Der Klang des Heckelphons". In: *Instrumentenbau-Zeitschrift* 20 (1966), p. 197.

[Mey66c]   J. Meyer. "Die Richtcharakteristiken von Oboen und Fagotten". In: *Das Musikinstrument* 15 (1966), p. 958.

[Mey67]   J. Meyer. "Die Richtcharakteristiken von Bratschen und Kontrabässen". In: *Instrumentenbau-Zeitschrift* 21 (1967), 3 and 116.

[Mey78]   J. Meyer. *Acoustics and the performance of music*. Frankfurt/Main: Verlag Das Musikinstrument, 1978.

[Mid99]   J. C. Middlebrooks. "Virtual localization improved by scaling non-individualized external-ear transfer functions in frequency". In: *The Journal of the Acoustical Society of America* 106 (1999), p. 1493.

[MT11]   M. Müller-Trapet, M. Pollow, and M. Vorländer. "Spherical harmonics as a basis for quantifying scattering and diffusing objects". In: *Proceedings of Forum Acusticum*. Aalborg, Denmark, 2011.

[Mül01]   S. Müller and P. Massarani. "Transfer-Function Measurement with Sweeps". In: *Journal of the Acoustical Society of America* (2001).

[Mül99]   S. Müller. "Digitale Signalverarbeitung für Lautsprecher". PhD thesis. RWTH Aachen University, 1999.

[Møl95]   H. Møller, M. F. Sørensen, D. Hammershøi, and C. B. Jensen. "Head-related transfer functions of human subjects". In: *Journal of the Audio Engineering Society* 43.5 (1995), pp. 300–321.

[Nai97]   M. Nair, M. Hegland, and R. Anderssen. "The trade-off between regularity and stability in Tikhonov regularization". In: *Mathematics of Computation of the American Mathematical Society* 66.217 (1997), pp. 193–206.

[Ohm06]   J. Ohm and H. Lüke. *Signalübertragung: Grundlagen der digitalen und analogen Nachrichtenbertragungssysteme*. Springer, 2006.

[Ota03]   M. Otani and S. Ise. "A fast calculation method of the head-related transfer functions for multiple source points based on the boundary element method". In: *Acoustical science and technology* 24.5 (2003), pp. 259–266.

[Oto04]   F. Otondo and J. H. Rindel. "The Influence of the Directivity of Musical Instruments in a Room". In: *Acta acustica united with Acustica* 90 (2004), pp. 1178–1184.

[Ped12]   A. Pedrero, M. Pollow, P. Dietrich, G. Behler, M. Vorländer, C. Díaz, and A. Díaz. "Mozarabic Chant anechoic recordings for auralization purposes". In: *FIA 2012*. 2012.

[Pel12a]   S. Pelzer, M. Pollow, and M. Vorländer. "Auralization of a virtual orchestra using directivities of measured symphonic instruments". In: *Proceedings of Acoustics 2012*. Nantes, France, 2012.

[Pel12b]   S. Pelzer, M. Pollow, and M. Vorländer. "Continuous and exchangeable directivity patterns in room acoustic simulation". In: *38th German Annual Conference on Acoustics (DAGA)*. Darmstadt, Germany, 2012.

[Pen03]   J. Pendleton. "Euler angle geometry, helicity basis vectors, and the Wigner D-function addition theorem". In: *American Journal of Physics* 71 (2003), p. 1280.

[Pen55]   R. Penrose. "A generalized inverse for matrices". In: *Mathematical Proceedings of the Cambridge Philosophical Society* 51 (1955), pp. 406–413.

[Pie89]   A. D. Pierce. *Acoustics: an introduction to its physical principles and applications*. Acoustical Soc of America, 1989.

[Pol07]   M. Pollow. "Variable directivity of dodecahedron loudspeakers". Diploma thesis. RWTH Aachen University, 2007.

[Pol09a]   M. Pollow. "Measuring directivities of musical instruments for auralization". In: *Proceedings of NAG/DAGA, International Conference on Acoustics*. Rotterdam, 2009, pp. 1471–1473.

[Pol09b]   M. Pollow, G. K. Behler, and B. Masiero. "Measuring Directivities of Natural Sound Sources with a Spherical Microphone Array". In: *Ambisonics Symposium*. Graz, Austria, 2009.

[Pol09c]   M. Pollow and G. K. Behler. "Variable Directivity for Platonic Sound Sources Based on Spherical Harmonics Optimization". In: *Acta Acustica united with Acustica* 95.6 (2009), pp. 1082–1092.

[Pol10a]   M. Pollow. "Multichannel dodecahedron loudspeakers for source directivity control". In: *EAA Euroregio, 1st European Congress on Sound and Vibration, 15 - 18 September 2010 : integrating the First Forum of Young Researchers in Acoustics 'EAA Summer School for Young Researchers', Ljubljana, Slovenia*. Vol. 96. 1. Stuttgart, 2010.

[Pol10b]   M. Pollow, G. Behler, and F. Schultz. "Musical Instrument Recording for Building a Directivity Database". In: *36th German Annual Conference on Acoustics (DAGA)*. Berlin, Germany, 2010.

[Pol11a]   M. Pollow and M. Vorländer. "Deriving continuous HRTFs from discrete data points". In: *37th German Annual Conference on Acoustics (DAGA)*. Düsseldorf, Germany, 2011, pp. 641–642.

[Pol11b]   M. Pollow, P. Dietrich, M. Kunkemöller, and M. Vorländer. "Synthesis of room impulse responses for arbitrary source directivities using spherical harmonic decomposition". In: *IEEE Workshop on Applications of Signal Processing to Audio and Acoustics (WASPAA)*. New Paltz, NY, USA, 2011, pp. 301–304.

[Pol12a]   M. Pollow, K.-V. Nguyen, O. Warusfel, T. Carpentier, M. Müller-Trapet, M. Vorländer, and M. Noisternig. "Calculation of head-related transfer functions for arbitrary field points using spherical harmonics decomposition". In: *Acta Acustica united with Acustica* 98.1 (2012), pp. 72–82.

[Pol12b]   M. Pollow, B. Masiero, P. Dietrich, J. Fels, and M. Vorländer. "Fast Measurement System for Spatially Continuous Individual HRTFs". In: *Spatial Audio in Today's 3D World - AES 25th UK Conference.* 2012.

[Pol12c]   M. Pollow, P. Dietrich, B. Masiero, and M. Vorländer. "Modal sound field representation of HRTFs". In: *38th German Annual Conference on Acoustics (DAGA)*. Darmstadt, Germany, 2012.

[Pol12d]   M. Pollow, J. Klein, P. Dietrich, G. K. Behler, and M. Vorländer. "Optimized Spherical Sound Source for Room Reflection Analysis". In: *Proceedings of the International Workshop on Acoustic Signal Enhancement (IWAENC)*. 2012.

[Pol12e]   M. Pollow, G. K. Behler, and M. Vorländer. "Post-processing and center adjustment of measured directivity data of musical instruments". In: *Proceedings of Acoustics 2012*. Nantes, France, 2012.

[Pol13a]   M. Pollow, J. Klein, P. Dietrich, and M. Vorländer. "Including directivity patterns in room acoustical measurements". In: *Proceedings of International Congress on Acoustics*. Montreal, Québec, Canada, 2013.

[Pol13b]   M. Pollow, P. Dietrich, and M. Vorländer. "Room Impulse Responses of Rectangular Rooms for Sources and Receivers of Arbitrary Directivity". In: *40th Italian (AIA) Annual Conference on Acoustics and the 39th German Annual Conference on Acoustics (DAGA).* Merano, Italy, 2013.

[Pol14a]   M. Pollow, J. Klein, S. Zillekens, and J. Fels. "Analysis and Processing of Rapidly Measured Individual HRTFs for Auralization". In: *EAA Joint Symposium on Auralization and Ambisonics (Poster presentation without paper).* 2014.

[Pol14b]   M. Pollow and M. Vorländer. "Efficient quality assessment of spatial audio data of high resolution". In: *40th German Annual Conference on Acoustics (DAGA).* Oldenburg, Germany, 2014.

[Pät10]   J. Pätynen and T. Lokki. "Directivities of Symphony Orchestra Instruments". In: *Acta acustica united with Acustica* (2010).

[Raf04]   B. Rafaely. "Plane-wave decomposition of the sound field on a sphere by spherical convolution". In: *Journal of the Acoustical Society of America* 116 (2004), p. 2149.

[Raf05]   B. Rafaely. "Analysis and design of spherical microphone arrays". In: *Speech and Audio Processing, IEEE Transactions on* 13.1 (2005), pp. 135–143.

[Raf07]   B. Rafaely, B. Weiss, and E. Bachmat. "Spatial Aliasing in Spherical Microphone Arrays". In: *IEEE Transactions on Signal Processing.* Vol. 55. 3. 2007, pp. 1003–1010.

[Ric14]   J.-G. Richter, M. Pollow, F. Wefers, and J. Fels. "Spherical harmonics based HRTF datasets: Implementation and evaluation for real-time auralization". In: *Acta Acustica united with Acustica* 100 (2014), pp. 667–675.

[Sch11]   D. Schröder. "Physically Based Real-Time Auralization of Interactive Virtual Environments". PhD thesis. RWTH Aachen University, 2011.

[Sch93]   A. Schmitz. "Naturgetreue Wiedergabe kopfbezogener Schallaufnahmen über zwei Lautsprecher mit Hilfe eines Übersprechkompensators". PhD thesis. RWTH Aachen University, 1993.

[Sha14]   N. Shabtai, M. Pollow, and M. Vorländer. "Acoustic Centering for High-Order Directivity Sources". In: *40th German Annual Conference on Acoustics (DAGA).* Oldenburg, Germany, 2014.

[Sle04]   K. Slenczka. "Simulation natürlicher Schallquellen für die binaurale Synthese". Diploma thesis. RWTH Aachen University, 2004.

[Slo04]   I. H. Sloan and R. S. Womersley. "Extremal systems of points and numerical integration on the sphere". In: *Advances in Computational Mathematics* 21.1-2 (2004), pp. 107–125.

[Slo99]   I. H. Sloan and R. S. Womersley. "The uniform error of hyperinterpolation on the sphere". In: *Mathematical Research* 107 (1999), pp. 289–306.

[Str03]   G. Strang. *Introduction to Linear Algebra*. Wellesley-Cambridge Press, 2003.

[Tuk77]   J. W. Tukey. "Exploratory data analysis: Past, present and future". In: (1977).

[Van06]   J. Vanderkooy. "The Acoustic Center: A New Concept for Loudspeakers at Low Frequencies". In: *Audio Engineering Society Convention 121*. 2006.

[Var06]   P. Vary and R. Martin. *Digital Speech Transmission. Enhancement, Coding and Error Concealment*. John Wiley and Sons, 2006.

[Vig07]   M. C. Vigeant, L. M. Wang, and J. H. Rindel. "Investigations of multi-channel auralization technique for solo instruments and orchestra". In: *Proceedings of International Congress on Acoustics*. Madrid, Spain, 2007.

[Wan08]   L. M. Wang and M. C. Vigeant. "Evaluations of output from room acoustic computer modeling and auralization due to different sound source directionalities". In: *Applied Acoustics* 69.12 (2008), pp. 1281–1293.

[War04]   O. Warusfel, E. Corteel, N. Misdariis, and T. Caulkins. "Reproduction of sound source directivity for future audio applications". In: *Proc. International Congress on Acoustics*. 2004.

[Wef10]   F. Wefers. "OpenDAFF - A free, open-source software package for directional audio data". In: *36th German Annual Conference on Acoustics (DAGA)*. 2010.

[Wei97]   G. Weinreich. "Directional tone color". In: *Journal of the Acoustical Society of America* 101 (1997), p. 2338.

[Wil99]   E. G. Williams. *Fourier Acoustics. Sound Radiation and Nearfield Acoustical Holography*. Academic Press, 1999.

[Zie14]   H. Ziegelwanger and P. Majdak. "Modeling the direction-continuous time-of-arrival in head-related transfer functions". In: *The Journal of the Acoustical Society of America* 135.3 (2014), pp. 1278–1293.

[Zil14]   S. Zillekens. "Messung von individuellen HRTFs und Nachbereitung mittels Kugeloberflächenfunktionen". Diploma thesis. RWTH Aachen University, 2014.

[Zot06]   D. N. Zotkin, R. Duraiswami, E. Grassi, and N. A. Gumerov. "Fast head-related transfer function measurement via reciprocity". In: *The Journal of the Acoustical Society of America* 120.4 (2006), pp. 2202–2215.

[Zot09a] F. Zotter. "Analysis and Synthesis of Sound-Radiation with Spherical Arrays". PhD thesis. Graz, Austria: University of Music and Performing Arts, 2009.

[Zot09b] F. Zotter. "Sampling Strategies for Acoustic Holography/Holophony on the Sphere". In: *Fortschritte der Akustik*. 2009.

List of Figures

List of Figures

Acknowledgments

A reseach work like this is only possible with the help of many diligent hands and many bright heads.

Primarily I would like to thank Prof. Michael Vorländer for the opportunity to work and conduct my research in such an inspiring environment. The possibility to work in his lab has been greatly appreciated. I owe my gratitude also to Prof. Peter Vary for the thorough review of this work and the friendly discussions that have been truly helpful. This thesis would not exist if Gottfried Behler had not employed the author quite some time ago as student assistant and taught him about acoustical measurements and spherical array systems. The fruitful discussions over the years have been tremendously beneficial for the outcome of this work.

I am indepted to all my colleagues at ITA who supported me over the years. Special thanks goes to Pascal Dietrich for always asking the right questions, his deep analytic view and his vigor to create things. Thanks to all ITA-Toolbox developers for building such a tremendously useful tool in great collaboration. I would like to thank Johannes Klein for his immense dedication and for the best pictures of our experiments. Thanks to Markus Müller-Trapet for numerous numerical simulations, fruitfull disussions and all the sound events we have experienced together. Obrigado pelo Bruno Masiero for the fresh ideas on science and the great churrascos here and there. Thanks to Dirk Schröder, Sönke Pelzer and Frank Wefers for the support in implementing directivity patterns in numerical simulation software that lead to great synergies and interesting publications.

A special thanks goes to Uwe Schlömer and Rolf Kaldenback in the name of their teams for the manifestations of our research ideas to solid physical objects. I am thankful for the students contributing scientifically to my research, namely Julian Blum, Johannes Klein, Tobias Knüttel, Benedikt Krechel, Martin Kunkemöller, Emma Lopez, Ander G. Perez Palacios, David Pazen, Jan Richter, Manual A. Galván Salazaar and Stefan Zillekens.

Thanks to many helpful discussions and projects with various fellow scientists, such as Filippo Fazi, Ilan Ben Hagai, Markus Noisternig and especially Franz Zotter. I am greatful for the joint projects with the research groups of Prof.

Stefan Weinzierl at TU Berlin, Germany, and Antonio Pedrero at Polytechnic University of Madrid, Spain, that fueled this work with data. Thanks to IRCAM, Paris, France, Ben-Gurion University of the Negev, Israel, Federal University of Santa Catarina, Brazil, and the University of São Paulo, Brazil, for the highly productive collaborations.

I would like to thank my wife Léa for her loving support and her patience, especially after all these long nights spent on the desk. And last but not least I would like to express my deepest gratitude to my parents who always encouraged me to do my thing, thank you very much for the everlasting support. This thesis is dedicated to you.

Bisher erschienene Bände der Reihe

Aachener Beiträge zur Technischen Akustik

ISSN 1866-3052

Alle erschienenen Bücher können unter der angegebenen ISBN-Nummer direkt online
(http://www.logos-verlag.de) oder per Fax (030 - 42 85 10 92) beim Logos Verlag
Berlin bestellt werden.